Random Motions in Markov and Semi-Markov Random Environments 1

Series Editor
Nikolaos Limnios

Random Motions in Markov and Semi-Markov Random Environments 1

Homogeneous Random Motions and their Applications

Anatoliy Pogorui
Anatoliy Swishchuk
Ramón M. Rodríguez-Dagnino

iSTE

WILEY

First published 2021 in Great Britain and the United States by ISTE Ltd and John Wiley & Sons, Inc.

ISTE Ltd
27-37 St George's Road
London SW19 4EU
UK

www.iste.co.uk

John Wiley & Sons, Inc.
111 River Street
Hoboken, NJ 07030
USA

www.wiley.com

Library of Congress Control Number: 2020946634

British Library Cataloguing-in-Publication Data
A CIP record for this book is available from the British Library
ISBN 978-1-78630-547-3

Contents

Preface

Motion is an essential element in our daily life. Thoughts related to motion can be found in the ancient Greek philosophers; however, we have to look several centuries ahead for relevant mathematical models. Galileo Galilei, Isaac Newton and Johannes Kepler, between the years 1550–1650, made remarkable advances in the construction of mathematical models for deterministic motion. Further advances in this line were made by Leonhard Euler and William Rowan Hamilton, and, in 1778, Joseph-Louis Lagrange proposed a new formulation of classical mechanics. This new formulation is based on the optimization of energy functionals, and it allows us to solve more sophisticated motion problems in a more systematic manner. This new approach to mechanics is the basis for dealing with modern quantum mechanics and the physics of high-energy particles. Albert Einstein, in 1905, through his special theory of relativity, introduced fine corrections for high velocities, close to the maximum speed of light. All of these fundamental mathematical models deal with many sophisticated motions such as that of astronomical objects, satellites (natural and artificial), particles in intense electromagnetic fields, particles under gravitational forces, a better understanding of light, and so on. However, even though these might be very complicated problems, we should mention that all of them have deterministic paths of motion.

In 1827, the Scottish botanist Robert Brown described a special kind of *random* motion produced by the interaction of many particles, while looking at pollen of the plant *Clarkia pulchella* immersed in water, using a microscope, and it was recognized that this type of motion could not be fully explained by modeling the motion of each particle or molecule. Albert Einstein, 78 years later, in 1905, published a seminal paper where he modeled the motion of the pollen as being moved by individual water molecules. In this work, the diffusion equation was introduced as a convenient mathematical model for this random phenomenon. A related model in this direction was presented in 1906 by Marian Smoluchowski, and the experimental verification was done by Jean Baptiste Perrin in 1908.

A similar model for Brownian motion was proposed in 1900 by Louis Bachelier in his PhD thesis entitled *The Theory of Speculation*, where he presented a stochastic analysis for valuing stock options in financial markets. This novel application of a stochastic model faced criticism at the beginning, but Bachelier's instructor Henri Poincaré was in full support of this visionary idea. This fact shows a close relationship between models to explain random phenomena in physics (statistical mechanics) and in financial analysis, and also in many other areas.

A notable contribution of the American mathematician Norbert Wiener was to establish the mathematical foundations for Brownian motion, and for that reason it is also known as the Wiener process. Great mathematicians such as Paul Lévy, Andrey Kolmogorov and Kiyosi Itô, among many other brilliant experts in the new field of probability and stochastic processes, set the basis of these stochastic processes. For example, the famous Black–Scholes formula in financial markets is based on both diffusion processes and Itô's ideas.

In spite of its success in modeling many types of random motion and other random quantities, the Wiener process has some drawbacks when capturing the physics of many applications. For instance, the modulus of velocity is almost always infinite at any instant in time, it has a free path length of zero, the path function of a particle is almost surely non-differentiable at any given point and its Hausdorff dimension is equal to 1.5, i.e. the path function is fractal. However, the actual movement of a physical particle and the actual evolution of share prices are barely justified as fractal quantities. Taking into account these considerations, in this book we propose and develop other stochastic processes that are close to the actual physical behavior of random motion in many other situations. Instead of the diffusion process (Brownian motion), we consider telegraph processes, Poisson and Markov processes and renewal and semi-Markov processes.

Markov (and semi-Markov) processes are named after the Russian mathematician Andrey Markov, who introduced them in around 1906. These processes have the important property of changing states under certain rules, i.e. they allow for abrupt changes (or switching) in the random phenomenon. As a result, these models are more appropriate for capturing random jumps, alternate velocities after traveling a certain random distance, random environments through the formulation of random evolutions, random motion with random changes of direction, interaction of particles with non-zero free paths, reliability of storage systems, and so on.

In addition, we will model financial markets with Markov and semi-Markov volatilities as well as price covariance and correlation swaps. Numerical evaluations of variance, volatility, covariance and correlations swaps with semi-Markov volatility are also presented. The novelty of these results lies in the pricing of volatility swaps in the closed form, and the pricing of covariance and correlation swaps in a market with two risky assets.

Anatoliy POGORUI
Zhytomyr State University, Ukraine

Anatoliy SWISHCHUK
University of Calgary, Canada

Ramón M. RODRÍGUEZ-DAGNINO
Tecnologico de Monterrey, Mexico

October 2020

Acknowledgments

Anatoliy Pogorui was partially supported by the State Fund for Basic Research (Ministry of Education and Science of Ukraine 20.02.2017 letter no. 12).

Anatoliy Pogorui

I would like to thank NSERC for its continuing support, my research collaborators, and my current and former graduate students. I also give thanks to my family for their inspiration and unconditional support.

Anatoliy Swishchuk

I would like to thank Tecnologico de Monterrey for providing me the time and support for these research activities. I also appreciate the support given by Conacyt through the project no. SEP-CB-2015-01-256237. The time and lovely support given to me by my wife Saida, my three daughters Dunia, Melissa and R. Melina, as well as my son Ramón Martín, are invaluable.

Ramón M. Rodríguez-Dagnino

Introduction

I.1. Overview

The theory of dynamical systems is one of the fields of modern mathematics that is under intensive study. In the theory of stochastic processes, researchers are actively studying dynamical systems, operating under the influence of random factors. A good representative of such systems is the theory of random evolutions. The first results within this field were obtained by Goldstein (1951) and Kac (1974), who studied the movement of a particle on a line with a speed that changes its sign under the Poisson process. Subsequently, this process was called the telegraph process or the Goldstein–Kac process. Further developments of this theory have been presented in the works of Griego and Hersh (1969, 1971), Hersh and Pinsky (1972), and Hersh (1974, 2003), which gave a definition of stochastic evolutions in a general setting.

Important advances in the theory of stochastic evolutions have been made in the formulation of limit theorems and their refinement; these consist of obtaining asymptotic expansions. These problems are studied in the works of Korolyuk and Turbin (1993), Skorokhod (1989), Turbin (1972, 1981), Korolyuk and Swishchuk (1986), Korolyuk and Limnios (2009, 2005), Shurenkov (1989, 1986), Dorogovtsev (2007b), Anisimov (1977), Girko (1982), Hersh and Pinsky (1972), Papanicolaou (1971a,b), Kertz (1978), Watkins (1984, 1985), Balakrishnan *et al.* (1988), Yeleyko and Zhernovyi (2002), Pogorui (1989, 1994, 2009a) and Pogorui and Rodríguez-Dagnino (2006, 2010a).

Among the many methods in finding limiting theorems, we should mention a class that could be called an asymptotic average scheme. By using these methods, a semi-Markov evolution can be reduced to a random evolution in a lumping state space with Markov switching, and is thus studied as a Markov scheme. Most of the results in this field are presented in the following books and papers: Korolyuk and Swishchuk (1995b), Korolyuk and Korolyuk (1999), Korolyuk and Turbin (1993),

Korolyuk and Limnios (2005), Turbin (1972), Pogorui (2004a, 2012a), and Rodríguez-Said *et al.* (2007). Furthermore, this theory was also developed in the study of asymptotic expansions for functionals of random evolution in the phase averaging and diffusion approximation. This topic was the main subject of the following works: Korolyuk and Limnios (2009, 2004), Turbin (1981), Samoilenko (2005), Albeverio *et al.* (2009), Pogorui (2010a), Pogorui and Rodríguez-Dagnino (2010a), and Nischenko (2001).

The asymptotic average scheme has been applied to a semi-Markov evolution for the computation of the effectiveness of a multiphase system with a couple of storage units by Pogorui (2003, 2004a), Pogorui and Turbin (2002), and Rodríguez-Said *et al.* (2007).

Research on random evolutions has also been carried out by applying martingale methods. It seems that the founders of this approach were Stroock and Varadhan (1969, 1979), but further developments can be found in the works of Skorokhod (1989), Pinsky (1991), Korolyuk and Korolyuk (1999), Sviridenko (1989), Swishchuk (1989), Hersh and Papanicolaou (1972), Iksanov and Résler (2006), and Griego and Korzeniowski (1989).

In addition to the successful development of abstract stochastic evolutions in the decade 1980–1990, several scholars studied various generalizations of the Goldstein–Kac telegraph process to multidimensional spaces. In connection to this, the results of Gorostiza (1973), Gorostiza and Griego (1979), Orsingher (1985), Orsingher and Somella (2004), Turbin (1998), Orsingher and Ratanov (2002), Samoilenko (2001) and Lachal (2006) should be noted. In most of these works, the authors considered a finite number of directions of a particle movement and obtained differential equations for the probability density functions. In the works of Pogorui (2007) and Pogorui *et al.* (2014), the authors proposed a method for solving such equations by using monogenic functions, after associating a particular commutative algebra with them.

Masoliver *et al.* (1993) studied the telegraph process with reflecting or partially reflecting boundaries and its distribution in a fixed interval of time. In papers by Pogorui (2005, 2006), and Pogorui and Rodríguez-Dagnino (2006, 2010b), the authors also study the stationary distribution of some Markov and semi-Markov evolution with delaying boundaries.

De Gregorio *et al.* (2005), and Stadje and Zacks (2004) considered the generalizationof the telegraph process on a line, where there is a discrete set of particle velocities, then at Poisson epochs a velocity is chosen from this set.

Pogorui (2010b), Pogorui and Rodríguez-Dagnino (2009b) and Samoilenko (2002) investigated fading evolutions, where the velocity of a particle tends to zero as the number of switches grows at infinite.

Pogorui and Rodríguez-Dagnino (2005a) also studied a generalization of the telegraph process to the case of Erlang interarrivals between successive switches of particle velocities. For such processes, a differential equation for the probability density function (pdf) of the particle position on the line was obtained. In addition, in Pogorui (2011a) a method for solving such equations, by using monogenic functions associated with this equation on a commutative algebra, was developed.

Orsingher and De Gregorio (2007), Stadje (2007) and others studied the motion of a particle in multidimensional spaces with constant absolute velocity and directions uniformly distributed on a unit sphere that change at Poisson points. The authors obtained explicit formulas for the position of the particle distributions for two- and four-dimensional space and investigated the "explosive effect" for the pdf of the position of a particle that approaches the singularity sphere for the plane and the three-dimensional space.

In recent years, much attention has been paid to the Pearson random walk with Gamma distributed steps and associated walks of Pearson–Dirichlet type, whose steps have the Dirichlet distribution. Franceschetti (2007) developed explicit formulas for the conditional pdf of the position of a particle at any number of steps for a walk in $\mathbb{R}^n$ ($n = 1, 2$) with uniformly distributed directions and steps with the Dirichlet distribution with parameter $q = 1$. Beghin and Orsingher (2010a) obtained an expression for the conditional distribution of the position of a particle of the Pearson–Dirichlet walk with parameter $q = 2$ in the plane. Le Caér (2010, 2011) generalized these results to the case of the multidimensional Pearson–Dirichlet random walk with arbitrary parameter q, where the author introduces the concept of a "Hyperspherical Uniform" (HU) random walk. The HU walk is a motion, the endpoint distribution of which is identical to the distribution of the projection in the walk space of a point, with a position vector, randomly chosen on the surface of the unit hypersphere of some hyperspace in higher dimensions. By using properties of the HU random walk, Le Caër found walks for which the conditional probability density can be expressed in a closed form. We should also mention the recent papers by De Gregorio (2014) and Letac and Piccioni (2014), where they obtained a generalization and simplification of proofs of the results stated by Le Caër.

In all of the above-mentioned papers, the authors study the conditional distributions of the particle position at renewal epochs of the switching direction process. For a non-Markov switching process, it is possible to find the corresponding results. The mathematical technique is to replace the distribution of the non-Markov process by the Markov chain walk embedded in this process. By using this approach, Pogorui and Rodríguez-Dagnino (2011a) obtained a recursive expression for the conditional characteristic functions of a random walk with Erlang switching, considering a non-Markov switching process. Namely, they studied the changes in the conditional characteristic functions of the particle position, not only at instants of the direction changing, but at all Poisson times. Pogorui (2011a) studied the particular case of the Erlang distributed stay of the switching process in the states. Further results in this direction were obtained by Pogorui and Rodríguez-Dagnino (2012) where the authors also studied multidimensional random motion at random velocities. For some distributions of random velocity, they observed an "explosive effect" for the pdf of the position of a particle when it is approaching the singularity sphere in four-dimensional space.

Other directions of random walk theory, which have been studied intensively during recent years, are the fractal Brownian motion and the fractal generalization of the telegraph process. These processes have been studied by Qian *et al.* (1998), Cahoy (2007), Beghin and Orsingher (2010b), Orsingher and Beghin (2009), D'Ovidio *et al.* (2014), and others.

The set of particles with interaction, where each particle moves on a line according to a telegraph process, up to collision with another particle, was studied by Pogorui (2012b). During the collision, the particles exchange momentums. In this book, the author calculates the distribution of time of the first collision for two telegraph particles that started simultaneously from different points on a line and investigates the limit of this distribution under Kac's condition. The author also investigates the system of particles with Markov switching, which is bounded with reflecting boundaries. The distribution for the position of particles of the system in a fixed time was also obtained. The limiting properties of these distributions and an estimate of the number of collisions in the system with reflecting boundaries, as well as without them, are also studied. Such a system of particles can be interpreted as a model of one-dimensional gas and it is a kind of one-dimensional generalization of the deterministic models of gas, of the billiard type, that were studied by Kornfeld *et al.* (1982), for example. The velocity of particles in these models is considered to be finite. This is a major difference from systems where the position of a particle is described by a diffusion process, such as in Arratia flow. We should note that models with finite speeds of particles moving under the influence of forces of mutual attraction were studied by Sinai (1992), Lifshits and Shi (2005), Giraud (2001, 2005), Bertoin (2002) and Vysotsky (2008).

I.2. Description of the book

The book is divided into two volumes, each containing two parts. Part 1 of Volume 1 consists of basic concepts and methods developed for random evolutions. These methods are the elementary tools for the rest of the book, and they include many results in potential operators and the description of some techniques to find closed-form expressions in relevant applications.

Part 2 of Volume 1 comprises three chapters (3, 4 and 5) dealing with asymptotic results (Chapter 3) and applications ranging from random motion with different types of boundaries, reliability of storage systems, telegraph processes, an alternative formulation to the Black–Scholes formula in finance, fading evolutions, jump telegraph processes and estimation of the number of level crossings for telegraph processes (Chapters 4 and 5).

Part 1 of Volume 2 extends many of the results of the latter part of Volume 1 to higher dimensions and consists of two chapters (1 and 2). Chapter 1 has the importance of presenting novel results of the random motion of the realistic three-dimensional case that has barely been mentioned in the literature. Chapter 2 deals with the interaction of particles in Markov and semi-Markov media, a topic many researchers have a strong interest in.

Part 2 of Volume 2 discusses applications of Markov and semi-Markov motions in mathematical finance across three chapters (3, 4 and 5). It includes applications of the telegraph process in modeling a stock price dynamic (Chapter 3), pricing of variance, volatility, covariance and correlation swaps with Markov volatility (Chapter 4), and the same pricing swaps with semi-Markov volatilities (Chapter 5).

The following is a general overview of the chapters and sections of the book. Chapters 1 and 2 of Volume 1 review the literature on the topic of random evolutions and outline the main areas of research. Many of these auxiliary results are used throughout the book.

Section 1.1 outlines research directions on the theory of telegraph processes and their generalizations.

In section 1.2, we introduce the notion of the projector operator and the generalized inverse operator or potential for an invertible reduced operator used in perturbation theory for linear operators. In turn, this theory is often used in the study of the asymptotic distribution of probability for reaching a "hard to reach domain".

In section 1.3, we consider the notion of a semigroup of operators generated by a Markov process. We give the definitions of the infinitesimal operator, the stationary distribution and the potential of a Markov process. These concepts are used in Chapter 3 for the asymptotic analysis of large deviations of semi-Markov processes.

Section 1.4 provides a constructive definition of a semi-Markov process based on the concept of the Markov renewal process (MRP). The notion of the semi-Markov kernel, which is a key definition for MRP, is considered. For a semi-Markov process, we introduce some auxiliary processes, with which a semi-Markov process forms a two-component (or bivariate) Markov process, and for such a process the infinitesimal operator is presented.

In section 1.5, we consider the notion of a lumped Markov chain and describe a phase merging scheme.

Section 1.6 describes a stochastic switching process in Markov and semi-Markov environments. We define semigroup operators associated with this process and consider their infinitesimal operator. In addition, the concept of superposition of independent semi-Markov processes is considered.

In Chapter 2 of Volume 1 we introduce homogeneous random evolutions (HRE), the elementary definitions, classification and some examples. We also present the martingale characterization and an analogue of Dynkin's formula for HRE. Some other important topics covered in this chapter are limit theorems, weak convergence and diffusion approximations, which are useful for Part 2 of Volume 2.

In Chapter 3 of Volume 1 we consider the asymptotic distribution of a functional of the time for reaching "hard to reach" areas of the phase space by a semi-Markov process on the line.

Section 3.1 is devoted to the analysis of the asymptotic distribution of a functional related to the time to reach a level that is infinitely removed by a semi-Markov process on the set of natural numbers.

In section 3.2, we give asymptotic estimates for the distribution of residence times of the semi-Markov process in the set of states that expands when the condition of existence of the functional A is not fulfilled.

In section 3.3, we obtain the asymptotic expansion for the distribution of the first exit time from the extending subset of the phase space of the semi-Markov process embedded in the diffusion process.

In section 3.4, we obtain asymptotic expansions for the perturbed semigroups of operators of the respective three-variate Markov process (after the standard extension of the phase space of the perturbed random evolution $u_\varepsilon (t, x)$ in the semi-Markov media), provided that the evolution $u_\varepsilon (t, x)$ weakly converges to the diffusion process as $\varepsilon > 0$.

In section 3.5, we obtain asymptotic expansions under Kac's condition in the diffusion approximation for the distribution of a particle position, which performs a random walk in a multidimensional space with Markov switching.

Section 3.6 describes a novel financial formula as an alternative to the well-known Black–Scholes formula for modeling the dynamic behavior of stock markets. This new formula is based on the asymptotic expansion for the singularly perturbed random evolution in Markov media.

Chapter 4 of Volume 1 is devoted to the computation of the stationary distributions for random switched processes with reflecting boundaries in Markov and semi-Markov environments. These results are used for the calculation of the efficiency of inventory control systems with feedback, and for some reliability problems of systems that are modeled by Markov and semi-Markov evolution.

In section 4.1, we derive the stationary distribution of transport processes with delaying in the boundaries in Markov media. These results are applicable to the study of multiphase systems with several reservoirs.

Section 4.2 deals with the transport process with semi-Markov switching. We find the stationary measure for this process and it is described by the differential equation with phase space on the corresponding interval and constant vector field values that depend on the switching semi-Markov process with a finite set of states.

In section 4.3, we give examples of applications of this method for the calculation of the stationary distributions for random switched processes with reflecting boundaries. In particular, we compute the efficiency coefficient of a single and two-phase inventory control system with feedback.

In section 4.4, we apply random evolutions with delaying barriers to modeling the control of supply systems with feedback, by considering the semi-Markov switching process.

In Chapter 5 of Volume 1 we study various models of stochastic evolutions that generalize the Goldstein–Kac telegraph process and investigate their distributions.

Section 5.1 is devoted to one-dimensional semi-Markov evolutions in an Erlang environment. In section 5.1.1, we derive a hyperbolic differential equation of the telegraph type for the pdf of the position of a particle moving at finite velocity. The interarrival times between two successive changes of velocity have an Erlang distribution.

In section 5.1.2, a method of solution of this partial differential equation is developed by using monogenic functions associated with a finite-dimensional commutative algebra. Section 5.1.3 extends the results of section 5.1.2 to infinite-dimensional commutative algebras. In section 5.1.4, by using the methods from sections 5.1.2 and 5.1.3, we obtain the distribution of a one-dimensional random evolution in Erlang media.

In section 5.2, we find the distribution of limiting positions of a particle moving according to a fading evolution. We assume that the interarrival times are Erlang or uniform distributed for the switching process.

In section 5.3, differential and integral equations for jump random motions are introduced and some examples have been developed to illustrate the method.

Section 5.4 is devoted to the estimation of the number of level crossings by the telegraph process in Kac's condition.

In Chapter 1 of Volume 2 we study random motions in higher dimensions. In section 1.1, we study the random walk of a particle with constant absolute velocity, which changes its direction according to a uniform distribution on the unit sphere at the renewal epochs of a switching process.

Section 1.2 deals with random motion with uniformly distributed directions and random velocity for the motion of particles in one, two, three and four dimensions. Similar cases are considered in section 1.3 for the distribution of random motion at non-constant velocity in semi-Markov media, and in section 1.4 for Goldstein–Kac telegraph equations and random flights in higher dimensions, where the directions of the movement and the velocity change at the renewal epochs. The jump telegraph process in $\mathbb{R}^n$ is considered in section 1.5.

In Chapter 2 of Volume 2 we study a system of interacting particles with Markov and semi-Markov switching. Section 2.1 is devoted to an ideal gas model with finite velocity of molecules. In this section, we find the distribution of the first meeting time of two telegraph particles on the line, which started simultaneously from different points.

In section 2.1.2, we estimate the number of particle collisions. In section 2.1.4, we obtain an asymptotic estimate of the number of collisions of the system of telegraph particles when time goes to infinity. Unlike the previous section, the system has no boundaries and particles can move arbitrarily far away in a straight line. Section 2.2 is devoted to the generalization of results of section 2.1 to the case of semi-Markov switching processes. In section 2.2.1, we obtain a set of renewal-type equations for the Laplace transform of the first collision of two particles. In section 2.2.2, we consider an example regarding a particular case of semi-Markov switching of particle velocity, and we study the limiting properties of the distribution of the particle position. Section 2.2.3 is devoted to studying the case where the first time of collision of two particles has finite expectation. We should note that in all previous examples, including the Markov case, expectation of the first time of collision of two particles is infinite.

In Chapter 3 of Volume 2 the application of the telegraph process for option prices, as an alternative to the diffusion process in the Black-Scholes formula, is further studied through asymptotic estimation of the corresponding operators. Some numerical results and plots are also presented.

Pricing variance, volatility derivatives, covariance and correlation swaps for financial markets with Markov-modulated volatilities are the main topics of Chapter 4 of Volume 2. These results are extended to the case of semi-Markov modulated volatilities in Chapter 5 of Volume 2. Numerical results and some plots to illustrate these results are also presented in these last two chapters.

PART 1

Basic Methods

1

Preliminary Concepts

1.1. Introduction to random evolutions

Evolution in a random environment means that the system depends on the state of the environment, and this occurs in many real systems in nature. Similarly, if the evolution of the system does not affect the random environment, but the environment is described by a random process, say a Markov renewal process (MRP), then such systems are called stochastic. Some stochastic systems change their states abruptly; that is, in every state, the system spends a random holding time and then immediately transfers to another state. Such systems are called discrete-state systems. The simulation of discrete stochastic systems often uses jumping Markov and semi-Markov processes. However, in many applications stochastic systems are changing their states continuously or in a combination of continuous and discrete, with the result of it being ineffective to model such systems with jumping Markov or semi-Markov processes. A better modeling strategy for such systems is the notion of semi-Markov (Markov) evolution, which is given by two processes: the switching MRP, describing the random environment, and the switched process that describes the evolution of the system.

This book is aimed at studying both discrete systems, the models for which are Markov and semi-Markov processes, and continuous systems, which are simulated by random evolutions (RE).

One of the relevant areas in the study of semi-Markov processes is the theory of large deviations, in particular, the asymptotic analysis of some functionals associated with the hitting time of a "hard-to-achieve" level. Such results are useful in queuing theory and reliability theory (Korolyuk and Turbin 1982).

One of the earliest asymptotic analyses of semi-Markov processes was done by Korolyuk, where he generalized the Vishyk–Lyusternik algorithm (Vasileva and

Butuzov 1973, 1990; Vishik and Lusternik 1960) to study the distributions of large deviations. Basically, the method consists of representing a distribution as an asymptotic series with two types of terms. One type of these terms regularly depends on a small parameter of perturbations of embedded (in the semi-Markov process) Markov chain, and the other type of terms is of the kind of boundary layer functions. This method allows us to study the asymptotic behavior of the residence time of a semi-Markov process in a fixed set of states. By using the approach of Korolyuk, the asymptotic analysis of hitting time of a "hard-to-achieve" level by semi-Markov processes was carried out by Korolyuk *et al.* (1973), Korolyuk and Tadjiev (1977), Korolyuk and Borovskikh (1981), Korolyuk and Turbin (1993, 1982), and others. An alternative method for the asymptotic analysis of singularly perturbed semigroups of operators, which makes it possible to write the asymptotic expansion coefficients explicitly, bypassing the recursive calculations, is presented in the papers by Turbin (1981) and Pogorui (1990, 1992, 1994).

The research to find the limiting distributions and the asymptotic analysis of semi-Markov processes continues to be an active area in recent times. For instance, we should mention in this line the works of Foss (2007), Kabanov and Pergamenshchiskov (2003), Semenov (2008), Silvestrov (2004, 2007a,b), Soloviev (1993), and Yeleyko and Zhernovyi (2002). Limiting distributions for stationary processes and their application in the assessment of financial risks are studied by Novak (2007, 2011).

The adapted Vishyk–Lyusternik algorithm also finds its application in the study of asymptotic expansions for functionals of random evolution in the phase averaging and diffusion approximation. This topic is studied in the works of Albeverio *et al.* (2009), Korolyuk and Limnios (2009, 2005), Samoilenko (2005), Pogorui (2009a), and Pogorui and Rodríguez-Dagnino (2010a).

Another important approach in the study of REs is the convergence rate of random walks by using limit theorems based on the martingale nature of RE. This has been the topic of various books and articles (Pinsky 1991; Korolyuk 1993; Swishchuk 1989; Iksanov 2006; Iksanov and Résler 2006; Sviridenko 1989).

Stochastic processes with Markov and semi-Markov switching to simulate the motion of a particle in a finite-dimensional space are called RE. One of the first works along this line was a random evolution model describing the motion of a particle on a line at a constant absolute speed whose directions are switched by the Poisson process. Such an analysis was made by Goldstein (1951) and Kac (1951, 1974). This model was later called the telegraph process because the distribution of the motion of such particles is a solution of the telegraph equation. However, the telegraph process $\{x(t)\}$ does not have some properties, which makes it difficult to study it. For example, the telegraph process $x(t)$, in contrast to the Wiener process, is neither Markov nor a martingale. In particular, these "defects" complicate the study of systems of interacting particles, whose trajectories are described by the telegraph processes. For example, after a hard collision of a particle with another particle, its whole trajectory is not described by a telegraph process. Pogorui (2012b) studied a system of particles and specified a manner to overcome problems arising from the non-Markov trajectories of particles.

The connections of the telegraph process with the random motion of particles have raised to new research areas:

1) The generalization of telegraph processes on multidimensional spaces, especially on a plane with a finite number of symmetrical directions of a particle movement, which changes at random Poisson events. These studies were performed in the works of Orsingher (1985), Orsingher and Somella (2004) and Lachal (2006).

2) The telegraph processes with reflecting and partially reflecting boundaries and their distribution. This problem was studied by Masoliver *et al.* (1993). The stationary distribution of Markov and semi-Markov RE with delays in boundaries (or sticky boundaries) and their applications to evaluate the effectiveness of the multiphase inventory control systems with feedback was studied by Pogorui and Turbin (2002), Pogorui (2003, 2004a), and Pogorui and Rodríguez-Dagnino (2009b, 2010b).

3) The generalizations of telegraph processes on heterogeneous cases in which the absolute speed is variable or the distribution switching processes change over time. For example, the work of Stadje and Zacks (2004) is devoted to the generalization of the telegraph process to the case where the particle velocity is an independent and identically distributed random variable that changes at Poisson events. Di Crescenzo and Martinucci (2010) studied a generalized telegraph process with increasing parameters of the switching process. Samoilenko (2002), Pogorui (2009b, 2010b), and Pogorui and Rodríguez-Dagnino (2009b, 2005b) studied fading evolution, where speed goes to zero at an infinite increase in the number of switches. In the paper by Pogorui and Rodríguez-Dagnino (2005a), the authors study a generalization of the telegraph process on the case of general Erlang sojourn times of the switching process and obtain the hyperbolic partial differential equation for the distribution of this process. In addition, in Pogorui (2011a) and Pogorui *et al.* (2014) a method for solving such kind of differential equation was developed.

4) Multidimensional random motion with an infinite number of directions of motion, which changes its direction at renewal epochs of the switching process, has been developed in some recent works. Apparently, one of the first papers in this field was that by Stadje (2007), where motion in the plane with Poisson switching process is studied. Further works in this direction are by Orsingher and De Gregorio (2007), and Franceschetti (2007). Most of these works use the apparatus of characteristic functions to study the distribution of a particle in multidimensional spaces with constant absolute speed when at Poisson events the particle changes its direction to another uniformly distributed on a unit sphere. In these papers,

the authors obtained the distribution of the particle position in explicit form for dimensions $n = 2, 4, 6$.

Pogorui and Rodríguez-Dagnino (2012) study an isotropic random motion in multidimensional space with a random velocity that is more natural from the physics point of view. When we have random velocity, then an "explosive effect" for some distributions is observed.

Pogorui and Rodríguez-Dagnino (2011b, 2013) studied an isotropic random motion with gamma steps in higher dimensions. Le Caér (2010, 2011), De Gregorio and Orsingher (2012), Beghin and Orsingher (2010a), De Gregorio (2014), and Letac and Piccioni (2014) considered a random motion in $\mathbb{R}^n$ by suggesting that switching points $0 < \tau_1 < \tau_2 < \cdots < \tau_n < t$ have the Dirichlet distribution on the interval $[0, t]$ and changes of direction have the uniform distribution on a sphere, which change at epochs τ_i.

In all of the papers mentioned above, the authors study the conditional distributions of the particle position at renewal epochs of the switching direction process.

5) Recently, there has been an increasing interest in studying stochastic flows in a system of interacting particles, the so-called Arratia flow. The first results in this field were found by Arratia (1979), where the author investigated a system of Wiener particles on a line that coalesce in the collision and continue to move as one particle. Many authors have continued this line of research and they have made important contributions (Dorogovtsev 2007b,a, 2010; Dawson 1993; Le and Raimond 2004; Yu 1999; Konarovskii 2011).

Pogorui (2012b) investigates a system of interacting particles with Markov switching. In particular, the author finds the distribution of the first time collision of two telegraph particles that started simultaneously from different points on the line, and he found the limit of the distribution under the Kac condition. Based on this result, the author investigates the time of free path for a family of particles with elastic collision, and he studies the distributions of particles both with reflecting boundaries and without them, and the limiting properties of these distributions.

Since the semi-Markov REs appear as abstract models for a wide class of real stochastic systems (Korolyuk 1987, 1993; Korolyuk *et al.* 1973; Korolyuk and Swishchuk 1986; Korolyuk and Turbin 1982), tthe study of their properties is important not only from a theoretical point of view but also in practical problems. In papers by Pogorui and Turbin (2002), Pogorui (2003, 2004b), Rodríguez-Said *et al.* (2007, 2008), and Pogorui *et al.* (2006), the authors study the stationary distributions of random evolution and their application to calculate the efficiency of inventory (or reservoir) control systems with feedback. Semi-Markov evolutions also find a wide use in the modeling and study of stochastic processes in finance and queueing systems, for example, in the works of Korolyuk (1993), Korolyuk and Swishchuk (1995b), Mitra (1998), Pogorui and Rodríguez-Dagnino (2009a), Pogorui and

Rodríguez-Dagnino (2008), (Ratanov 2007), López and Ratanov (2012), Swishchuk and Burdeinyi (1996), Swishchuk (2004), Maglaras and Zeevi (2004), and others.

1.2. Abstract potential operators

In the study of the asymptotic distribution of probability of reaching a "hard-to-reach domain" by semi-Markov processes, the theory of perturbation for linear operators is systematically used. As basic tools for this theory, it is necessary to introduce the notion of the projector and the generalized inverse operator or potential operator.

Let $\mathcal{B}$ be a Banach space. Two closed linear manifolds M_1 and M_2 in $\mathcal{B}$ are called supplemented if $M_1 \cap M_2 = 0$, where 0 is the zero element in $\mathcal{B}$, and $M_1 + M_2 = \{x : x = x_1 + x_2,\ x_1 \in M_1,\ x_2 \in M_2\} = \mathcal{B}$. Then, we say that $\mathcal{B}$ is a direct sum of M_1 and M_2 and we use the usual notation $\mathcal{B} = M_1 \oplus M_2$. If M_1 and M_2 are closed subspaces in $\mathcal{B}$, then to each image in the form of a direct sum corresponds a limited projector P such that $P\mathcal{B} = M_1$, $(I - P)\,\mathcal{B} = M_2$.

Let $A : \mathcal{B} \to \mathcal{B}$ be a linear operator, and we have the following definitions:

DEFINITION 1.1.– (Kato 1980) *A linear operator A, defined on some linear manifold $\mathcal{D}(A)$, is called closed if from $x_n \to x$ $(x_n \in \mathcal{D}(A))$ and $Ax_n \to y$ it follows that $x \in \mathcal{D}(A)$ and $y = Ax$.*

DEFINITION 1.2.– (Kato 1980) *A closed densely defined operator A $\left(\overline{\mathcal{D}(A)} = \mathcal{B}\right)$ is called normally solvable if its range $R(A)$ is a subspace of $\mathcal{B}$.*

The following theorem characterizes normally solvable operators:

THEOREM 1.1.– (Kato 1980) A closed densely defined operator A, with $R(A) \neq \mathcal{B}$, is normally solvable if and only if (iff)

$$N(A^*)^{\perp} = R(A),$$

where $N(A^*)^{\perp}$ is a subset of $\mathcal{B}$ such that $f(x) = 0$ for all $f \in N(A^*)$.

The following lemma states another important condition of normal solvability of operators:

LEMMA 1.1.– (Kato 1980; Krein 1971) Suppose A is a closed operator and there exists a closed linear subspace M in $\mathcal{B}$ such that $\mathcal{B} = M \oplus R(A)$. Then the operator A is normally solvable.

The following characteristic for normal solvability of an operator is well known (Cox 1976). A bounded linear operator A is a normally solvable operator if and only if its restriction $\overline{A}$ on $R(A)$ has a bounded inverse operator $\overline{A}^{-1}$.

DEFINITION 1.3.– (Kato 1980) *A normally solvable operator A is called a Fredholm operator if* $\dim N(A^*) = \dim N(A) = r < \infty$.

DEFINITION 1.4.– (Kato 1980; Krein 1971) *A closed densely defined operator A on $\mathcal{B}$ is called reducible-invertible if $\mathcal{B}$ can be represented as follows:*

$$\mathcal{B} = N(A) \oplus R(A). \qquad [1.1]$$

Since $R(A)$ is closed, then a reducible-invertible operator is normally solvable. Hence, for a reducible-invertible operator A, we have $R(A) \cap N(A) = \{0\}$ and $\forall b \in \mathcal{B}$ can be represented in the form $b = f + u$, where $f \in R(A)$, $u \in N(A)$ and $N(A)$ and $R(A)$ are subspaces of $\mathcal{B}$.

The decomposition in equation [1.1] generates a projective operator Π on $N(A)$ such that

$$\Pi N(A) = N(A), \qquad (I - \Pi) R(A) = R(A). \qquad [1.2]$$

DEFINITION 1.5.– (Kato 1980) *A projective operator Π, which satisfies properties of equation* [1.2] *for some reducible-invertible operator A, is said to be the proper projector of A.*

The proper projector Π satisfies the following conditions $\forall \varphi \in B$:

1) $\Pi^2 \varphi = \Pi \varphi$;

2) $\Pi A \varphi = A \Pi \varphi = 0$.

LEMMA 1.2.– (Kato 1980; Korolyuk and Turbin 1993) If A is a reducible-invertible operator, then there exists a bounded inverse operator $(A + \Pi)^{-1}$.

Let us consider an example of a proper projective operator of a reducible-invertible operator A. Let E be a countable set. Denote by l_∞ the real space (Banach space) of vectors $\mathbf{u} = \{u_i,\ i \in E\}$ with the norm $\|\mathbf{u}\| = \max_{i \in E} |u_i|$ and by l_1 the real space of vectors $\boldsymbol{\rho} = \{\rho_i,\ i \in E\}$ with the norm $\|\boldsymbol{\rho}\| = \sum_{i \in E} |\rho_i|$. For any $\boldsymbol{\rho} \in l_1$ and $\mathbf{u} \in l_\infty$, define the scalar product

$$(\boldsymbol{\rho}, \mathbf{u}) = \sum_{i \in E} \rho_i u_i.$$

The tensor product of $\mathbf{u} \in l_\infty$ and $\boldsymbol{\rho} \in l_1$ is defined as follows:

$$[\mathbf{u} \otimes \boldsymbol{\rho}] = \{u_i \rho_j,\ i, j \in E\}.$$

It is easy to see that for $\mathbf{f} \in l_\infty$, we have

$$[\mathbf{u} \otimes \boldsymbol{\rho}]\, \mathbf{f} = (\boldsymbol{\rho}, \mathbf{f})\, \mathbf{u}.$$

Now, suppose that an operator $A = \{a_{ij},\ i, j \in E\}$ with real a_{ij} satisfies the condition

$$\sup_{i \in E} \sum_{j \in E} |a_{ij}| = \|A\| < \infty. \qquad [1.3]$$

Consider

$$A\mathbf{u} = \sum_{j \in E} a_{ij} u_j.$$

It follows from equation [1.3] that $A : l_\infty \to l_\infty$. In addition, A is bounded on l_∞ with the norm $\|A\|$.

The adjoint operator A^* of A is defined on l_1 as follows:

$$A^* \boldsymbol{\rho} = \sum_{i \in E} a_{ij} \rho_i.$$

It is easily verified that A^* is bounded on l_1 and $\|A^*\| = \|A\|$.

Suppose A is Fredholm, i.e. $\dim N(A) = \dim N(A^*) = r < \infty$, where $N(A)$ and $N(A^*)$ are the kernels of operators A and A^*, respectively.

Let $u^{(1)}, u^{(2)}, \dots, u^{(r)}$ be a basis of $N(A)$ and $\rho^{(1)}, \rho^{(2)}, \dots, \rho^{(r)}$ be a basis of $N(A^*)$. Suppose that

$$\left(\rho^{(i)}, u^{(j)}\right) = \delta_{ij},$$

where δ_{ij} is the Kronecker symbol.

The operator $\Pi = \sum_{k=1}^{r} \left[u^{(k)} \otimes \rho^{(k)}\right]$ is the proper operator of A and it satisfies the following properties:

1) $\Pi^2 = \Pi$;

2) $\Pi u = u$, $u \in N(A)$;

3) $\rho \Pi = \rho$, $\rho \in N(A^*)$;

4) $\Pi f = 0$, $f \in R(A)$;

5) $\varphi \Pi = 0$, $\varphi \in R(A^*)$;

6) $\Pi A \varphi = A \Pi \varphi$, $\varphi \in R(A)$.

LEMMA 1.3.– (Kato 1980; Korolyuk and Turbin 1993) Suppose that an operator A is reducible-invertible and Π is its proper projector. Then, there exists the bounded inverse operator $(A + \Pi)^{-1}$.

It follows from lemma 1.3 that there exists $(A+\Pi)^{-1}-\Pi$ and it is bounded.

DEFINITION 1.6.– (Korolyuk and Turbin 1976, 1993; Sato 1971) *The operator* $R_0 = (A+\Pi)^{-1}-\Pi$ *is called the generalized inverse operator or potential operator of* A.

The operator R_0 *is also called the abstract potential operator of the reducible-invertible operator* A.

LEMMA 1.4.– (Kato 1980; Korolyuk and Turbin 1993) The operator R_0 satisfies the following properties:

$$AR_0 = R_0A = I - \Pi;$$

$$R_0\Pi = \Pi R_0 = 0;$$

$$\|R_0\| = \left\|\overline{A}^{-1}\right\|,$$

where $\overline{A}$ is the restriction of A to $R(A)$.

The notion of a generalized inverse operator plays an important role in the study of Markov processes (Gikhman and Skorokhod 1975; Korolyuk and Turbin 1993, 1982). In particular, for a Markov chain with transition probabilities matrix P, the generalized inverse operator R_0 of $I-P$ is called the potential of this Markov chain.

EXAMPLE 1.1.– Consider a Markov chain $\{\xi_n,\, n \geq 0\}$ with transition probabilities matrix

$$P = \begin{pmatrix} 1 & 0 & 0 & 0 \\ 1 & 0 & 0 & 0 \\ 0 & 0 & \frac{1}{4} & \frac{3}{4} \\ 0 & 0 & \frac{1}{2} & \frac{1}{2} \end{pmatrix}.$$

Denote by A the following matrix

$$A = I - P = \begin{pmatrix} 0 & 0 & 0 & 0 \\ -1 & 1 & 0 & 0 \\ 0 & 0 & \frac{3}{4} & -\frac{3}{4} \\ 0 & 0 & -\frac{1}{2} & \frac{1}{2} \end{pmatrix}.$$

The matrix A is called infinitesimal for the Markov chain ξ_n. It is easily verified that $u^{(1)} = \begin{pmatrix} 1 \\ 1 \\ 0 \\ 0 \end{pmatrix}$, $u^{(2)} = \begin{pmatrix} 0 \\ 0 \\ 1 \\ 1 \end{pmatrix}$ is a basis of $N(A)$ and $\rho^{(1)} = \left(1,\, 0,\, 0,\, 0\right)$, $\rho^{(2)} = \left(0,\, 0,\, \frac{2}{5},\, \frac{3}{5}\right)$ is a basis of $N(A^*)$, where A^* is the transpose matrix of A. Thus, $\dim N(A) = \dim N(A^*)$.

Let us calculate the following scalar and tensor products $\left(\rho^{(i)}, u^{(j)}\right) = \delta_{ij}$,

$$\left[u^{(1)} \otimes \rho^{(1)}\right] = \begin{pmatrix} 1 \\ 1 \\ 0 \\ 0 \end{pmatrix} (1,\ 0,\ 0,\ 0,) = \begin{pmatrix} 1\ 0\ 0\ 0 \\ 1\ 0\ 0\ 0 \\ 0\ 0\ 0\ 0 \\ 0\ 0\ 0\ 0 \end{pmatrix},$$

$$\left[u^{(2)} \otimes \rho^{(2)}\right] = \begin{pmatrix} 0 \\ 0 \\ 1 \\ 1 \end{pmatrix} (0,\ 0,\ \tfrac{2}{5},\ \tfrac{3}{5}) = \begin{pmatrix} 0\ 0\ 0\ 0 \\ 0\ 0\ 0\ 0 \\ 0\ 0\ \frac{2}{5}\ \frac{3}{5} \\ 0\ 0\ \frac{2}{5}\ \frac{3}{5} \end{pmatrix}.$$

Thus, the proper projector of A is as follows:

$$\Pi = \begin{pmatrix} 1\ 0\ 0\ 0 \\ 1\ 0\ 0\ 0 \\ 0\ 0\ \frac{2}{5}\ \frac{3}{5} \\ 0\ 0\ \frac{2}{5}\ \frac{3}{5} \end{pmatrix}.$$

Then,

$$A + \Pi = \begin{pmatrix} 1 & 0 & 0 & 0 \\ 0 & 1 & 0 & 0 \\ 0 & 0 & \frac{23}{20} & -\frac{3}{20} \\ 0 & 0 & -\frac{1}{10} & \frac{11}{10} \end{pmatrix}.$$

Therefore,

$$(A + \Pi)^{-1} = \begin{pmatrix} 1 & 0 & 0 & 0 \\ 0 & 1 & 0 & 0 \\ 0 & 0 & \frac{22}{25} & \frac{3}{25} \\ 0 & 0 & \frac{2}{25} & \frac{23}{25} \end{pmatrix}.$$

From this we have the potential of Markov chain ξ_n in the form of the generalized inverse operator A:

$$R_0 = (A + \Pi)^{-1} - \Pi = \begin{pmatrix} 0 & 0 & 0 & 0 \\ -1 & 1 & 0 & 0 \\ 0 & 0 & \frac{12}{25} & -\frac{12}{25} \\ 0 & 0 & -\frac{8}{25} & \frac{8}{25} \end{pmatrix}.$$

1.3. Markov processes: operator semigroups

In this section, we study elementary concepts of semigroup of operators generated by Markov process. We also define the infinitesimal operator, and the stationary distribution of Markov processes. These notions will be used for the asymptotic analysis (large deviations) of semi-Markov processes.

Denote by $(\Omega, \mathcal{F}, \mathcal{P})$ a probability space and by (X, Σ) a complete separable metric space, where Σ is a σ-algebra of Borel subsets of X.

Let $\{\xi(t),\, t \geq 0\}$ be a homogeneous Markov process on $(\Omega, \mathcal{F}, \mathcal{P})$ on the phase space (X, Σ). Denote by $P(t, x, B) = P\{\xi(t) \in B | \xi(0) = x\}$ the transition probability of $\xi(t)$. It is well known that for every $t \in T$ the probability $P(t, x, B)$ is a stochastic kernel, i.e. for a fixed $x \in X$, $P(t, x, B)$ is a measure on Σ and $P(t, x, X) = 1$, and for a fixed B, $P(t, x, B)$ is a measurable function with respect to x.

Denote by $B(\Sigma)$ a set of bounded and Σ-measurable functions on X. Consider the family of operators generated by the transition probabilities: for a function $f \in B(\Sigma)$, we define

$$T_t f(x) = \int_X f(y)\, P(t, x, dy).$$

It is easy to see that $T_t : B(\Sigma) \to B(\Sigma)$, and it follows from the Chapman–Kolmogorov equation that

$$T_{s+t} = T_s\, T_t.$$

Thus, the family of operators T_t, $t \in T$ is a semigroup.

Let D be a subset of $B(\Sigma)$ such that there exist the following limits for any $\varphi \in D$

$$A\varphi = \lim_{\Delta t \to 0+} \frac{T_{\Delta t}\varphi - \varphi}{\Delta t},$$

$$\lim_{\Delta t \to 0+} T_{\Delta t}\varphi = \varphi.$$

The operator A is called the infinitesimal operator of the semigroup of operators T_t and of the Markov process $\xi(t)$ (Gikhman and Skorokhod 1975; Kovalenko 1980; Kovalenko *et al.* 1983), respectively.

DEFINITION 1.7.– *It is said that $\xi(t)$ has the stationary distribution π if for any $B \in \Sigma$*

$$\pi(B) = \int_X \pi(dx) P(t, x, B), \quad t \geq 0.$$

The stationary projector Π of $\xi(t)$ is defined as follows:

$$\Pi f(x) = \int_X \pi(ds) f(s) \mathbb{I}(x),$$

where $\mathbb{I}(x) = 1$ for all $x \in X$ and 0 otherwise.

DEFINITION 1.8.– *A Markov process $\xi(t)$ is said to be uniformly ergodic if $\|T_t - \Pi\| \to 0$ as $t \to \infty$.*

It is well known (Korolyuk and Turbin 1976, 1993) that the infinitesimal operator A of a uniformly ergodic Markov process $\xi(t)$, $t \geq 0$ is reducible-invertible. Thus, there exists the inverse operator $(A + \Pi)^{-1}$. Moreover, the operator $R_0 = (A + \Pi)^{-1} - \Pi$ is called the potential of the stochastic process $\xi(t)$.

Let us define $Q = I - \Pi$. Then, the potential R_0 satisfies the following properties (Korolyuk and Turbin 1976, 1993):

1) $R_0\Pi = \Pi R_0 = 0$;

2) $R_0 Q = QR_0 = R_0$;

3) $R_0 A = AR_0 = Q$.

DEFINITION 1.9.– *The number $\varpi = \lim_{t\to\infty} \frac{1}{t} \ln \|T_t\|$ is called the type of a semigroup T_t.*

The following property of the resolvent $R(\lambda, A)$ of operator A is important (Korolyuk and Turbin 1993):

THEOREM 1.2.– If semigroup T_t is of type ϖ, then for any λ such that $\operatorname{Re}\lambda > \varpi$ we have

$$R(\lambda, A) f := (\lambda I - A)^{-1} f = \int_0^\infty e^{-\lambda t} T(t) f dt, \quad f \in B(\Sigma).$$

The resolvent of a reducible-invertible operator A has the following representation for $0 < |\lambda| < \frac{1}{R_0}$ (Korolyuk and Turbin 1993, 1982):

$$R(\lambda, A) = \frac{1}{\lambda}\Pi + R_0(\lambda R_0 - I)^{-1}.$$

Hence, it follows that for a reducible-invertible operator A for $0 < |\lambda| < \frac{1}{R_0}$

$$\Pi R(\lambda, A) = \frac{1}{\lambda}\Pi,$$

$$QR(\lambda, A) = R_0(\lambda R_0 - I)^{-1}.$$

Now, let us define $T_t - \Pi = \mathfrak{H}(t)$. For a uniformly ergodic Markov process, the integral $\int_0^\infty \mathfrak{H}(t)dt$ converges and (Sato 1971):

$$\lim_{\lambda \downarrow 0} \int_0^\infty e^{-\lambda t} \mathfrak{H}(t)dt = \int_0^\infty \mathfrak{H}(t)dt.$$

It can be proved (Kato 1980; Korolyuk and Turbin 1993) that if the following two conditions are fulfilled, i.e. $\forall f \in D(R_0)$, $\int_0^{+\infty} \|(T_t - \Pi) f\| \, dt < +\infty$, then the following equality holds:

$$\int_0^{+\infty} [T_t - \Pi] \, f dt = \int_0^\infty \mathfrak{H}(t) \, f dt = -R_0 f.$$

1.4. Semi-Markov processes

This section is devoted to providing a constructive definition of a semi-Markov process, which is based upon the notion of an MRP, and a semi-Markov kernel, which is a crucial part of the definition of a semi-Markov process. In addition, we introduce an auxiliary process useful in reducing a semi-Markov process to the Markov with the standard phase space extension. For this Markov process, the infinitesimal operator is presented.

DEFINITION 1.10.– *A semi-Markov kernel in a measurable space (X, Σ) is defined by a function (x, B, t) $x \in X$, $B \in \Sigma$, $t \in [0, +\infty]$, which satisfies the following conditions:*

1) $Q(x, B, t)$ *is non-decreasing and right-continuous with respect to* $t \geq 0$ *for fixed* $x \in X$ *and* $B \in \Sigma$, and $Q(x, B, 0) = 0$ *for* $x \notin B$;

2) *for a fixed* $t \geq 0$, $Q(x, B, t)$ *is a semi-stochastic kernel, i.e.* $\forall x \in X$ $Q(x, X, t) \leq 1$;

3) $Q(x, B, +\infty)$ *is a stochastic kernel with respect to* x *and* B, *i.e.* $Q(x, X, +\infty) = 1$.

DEFINITION 1.11.– *A two-component Markov chain $\{\xi_n, \theta_n;\ n \geq 0\}$ on the phase space $X \times [0, +\infty)$ is said to be an MRP if its transition probabilities depend only on the value of the first component ξ_n and is defined by a semi-Markov kernel $Q(x, B, t)$ as follows:*

$$P\{\xi_{n+1} \in B, \theta_{n+1} \leq t | \, \xi_n = x\} = Q(x, B, t).$$

The first component $\{\xi_n,\ n \geq 0\}$ of $\{\xi_n, \theta_n;\ n \geq 0\}$ is a Markov chain with the transition probabilities $P(x, B) = Q(x, B, +\infty)$ and it is called the embedded Markov chain.

The non-negative random variables θ_n, $n \geq 0$ in $\{\xi_n, \theta_n;\ n \geq 0\}$ are the sojourn times of the MRP

$$\tau_n = \sum_{k=1}^{n} \theta_k, \quad n \geq 1, \quad \tau_0 = \theta_0 = 0.$$

The distribution function of θ_{n+1} depends on the state ξ_n as follows:

$$G_x(t) = P\{\theta_{n+1} \leq t |\, \xi_n = x\} = Q(x, X, t).$$

A random variable θ_{n+1} with the distribution function $G_x(t)$ is interpreted as the holding time of the MRP in a state $\xi_n = x$ and it is convenient to denote it by θ_x.

Thus, to complete the constructive definition of the MRP, it is necessary to define a semi-Markov kernel.

Since for all $t \geq 0$ and $x \in X$, $Q(x, B, t) \leq P(x, B)\ \forall B \in \Sigma$, the measure Q is absolutely continuous with respect to P. Consequently, there exists a measurable function $G(x, B, t)$ such that

$$Q(x, B, t) = \int_B G(x, y, t)\, P(x, dy).$$

The function $G(x, y, t)$ is a conditional distribution of θ_{n+1} assuming that the embedded Markov chain transits from x to y:

$$G(x, y, t) = P\{\theta_{n+1} \leq t |\, \xi_n = x,\, \xi_{n+1} = y\}.$$

For a fixed trajectory of the Markov chain $\{\xi_n,\ n \geq 0\}$, the random variables θ_n, $n \geq 0$ are independent and it can be said that they are conditionally independent. Indeed, by using the formula of total probability, we have

$$P\{\theta_1 \leq t_1, \cdots, \theta_{n+1} \leq t_{n+1} |\, \xi_1 = x_1, \cdots, \xi_n = x_n,\, \xi_{n+1} = x_{n+1}\}$$
$$= \prod_{k=0}^{n} G(x_k, x_{k+1}, t_{k+1}). \qquad [1.4]$$

The conditional independence of θ_n, $n \geq 0$ for a fixed trajectory of the Markov chain $\{\xi_n,\ n \geq 0\}$ makes it possible to give another definition of the MRP $\{\xi_n,\ \theta_n,\ n \geq 0\}$, which is based on a sequence of non-negative independent random variables $\{\theta_n,\ n \geq 0\}$. These random variables are defined on the Markov chain $\{\xi_n,\ n \geq 0\}$ by the joint distribution [1.4].

It was shown in Korolyuk and Turbin (1976) and Silvestrov (1980) that if X is a countable set then, without loss of generality, we may assume that $G(x, y, t)$ does not depend on y, i.e. the semi-Markov kernel is of the following form:

$$Q(x, B, t) = P(x, B)\, G_x(t),$$

where $G_x(t) = P(\theta_x \leq t |\, \xi_n = x)$ and θ_x is a sojourn time of the MRP at state x.

Let us introduce a counting process $\{\nu(t)\}$ as follows:

$$\nu(t) = \sup\{n : \tau_n \leq t\}.$$

The process $\nu(t)$ counts the number of renewal occurrences in the interval $[0, t]$.

LEMMA 1.5.– The process $\xi(t) := \xi_{\nu(t)}$, $t \geq 0$, is semi-Markov.

If, in addition, for every $t \geq 0$, $P\{\nu(t) < \infty\} = 1$, then the process $\xi_{\nu(t)}$ is called regular.

For a semi-Markov process $\xi(t)$ define

$$\tau(t) = t - \sup\{u \leq t : \xi(u) \neq \xi(t)\}.$$

The process $\tau(t)$ is called an auxiliary process. Let us consider the bivariate process $\zeta(t) = (\xi(t), \tau(t))$ on the phase space $X \times [0, +\infty)$. It is well known that the process $\zeta(t)$ is a Markov process (Gikhman and Skorokhod 1975; Corlat *et al.* 1991; Korolyuk and Turbin 1982).

Let $\varphi(x, t)$, $x \in X$, $t \in [0, +\infty)$ be a $\Sigma \times \mathcal{R}_+$-measurable function differentiable with respect to t. Then the infinitesimal operator of $\zeta(t)$ is of the following form (Corlat *et al.* 1991; Korolyuk and Turbin 1982):

$$A\varphi(x, t) = \frac{d}{dt}\varphi(x, t) + \frac{g_x(t)}{1 - G_x(t)}\left[P\varphi(x, 0) - \varphi(x, t)\right],$$

where $G_x(t) = P(\theta_x < t)$, $g_x(t) = \frac{d}{dt}G_x(t)$ and $P(x, dy)$ is the transition probability of the Markov chain $\{\xi_n,\ n \geq 0\}$. In addition,

$$P\varphi(x, 0) = \int_X P(x, dy)\varphi(y, 0).$$

A jumping Markov process is a particular case of a semi-Markov process with the following semi-Markov kernel:

$$Q(x, B, t) = P(x, B)\left(1 - e^{-q(x)t}\right),\ q(x) > 0,\ t \geq 0.$$

The function $q(x),\ x \in X$ determines the intensity of staying at a particular state x. The infinitesimal operator of the jumping Markov process is given as (Korolyuk 1993; Korolyuk and Korolyuk 1999):

$$A\varphi(x) = q(x)\left[\int_X \varphi(y)P(x, dy) - \varphi(x)\right],\ x \in X,\ B \in \Sigma,$$

where φ is a real Σ-measurable function, which is bounded on X.

1.5. Lumped Markov chains

We will describe a phase merging or state lumping scheme for Markov chains. This scheme applies to the investigation of the probability distribution of reaching a "hard-to-reach domain" by semi-Markov processes. This state lumping scheme was developed and introduced in seminal works by Korolyuk and Turbin (1976, 1982). In Korolyuk *et al.* (1979a,b), lumping states of the phase space of the Markov chain are described, and we will describe the basic issues of it in order to use it as an approach for the investigation of the probability distribution of reaching a level, which goes to infinity.

We will describe the state lumping algorithm for Markov chains.

Let $\{\xi_n,\ n \geq 0\}$ be a homogeneous Markov chain with the phase or state space $E = \{1, 2, \ldots, M\}$ and the matrix of transition probabilities $P = \{p_{ij},\ i, j \in E\}$. We assume that all states of the Markov chain communicate, i.e. they are accessible to each other or $\forall\, i, j \in E$, $\exists\, l \geq 1$, such that $p_{ij}^{(l)} > 0$.

Let us establish that $E_0 = \{1, 2, \ldots, m\}$, $m < M$, $E_1 = E \setminus E_0$. Consider the following random time instants:

$$\begin{cases} \nu_0 = 0, \\ \nu_1 = \min\{n > 0 : \xi_n \in E_0\} \\ \vdots \\ \nu_k = \min\{n > \nu_{k-1} : \xi_n \in E_0\}. \end{cases}$$

Now, we introduce the sequence

$$\xi_k^{E_0} = \xi_{\nu_k},\ k \geq 0,\ \xi_0^{E_0} = \xi_0 \in E_1.$$

It is easily verified that $\xi_k^{E_0}$, $k \geq 0$ is a homogeneous Markov chain with the state space $E_0 = \{1, 2, \ldots, m\}$ and transition probabilities matrix $P^{E_0} = \left\{p_{ij}^{E_0},\ i, j \in E_0\right\}$. Moreover

$$P^{E_0} = P_{00} + P_{01}(I - P_{11})^{-1} P_{10},$$

where P_{kl}, k, $l = 0, 1$, are matrices with elements p_{ij}, $i \in E_k$, $j \in E_l$.

Since the matrix P_{11} is not stochastic by assumption, then there exists the inverse matrix $(I - P_{11})^{-1} = \sum_{k=0}^{\infty} P_{11}^k < \infty$ (Kemeny and Snell 1960).

DEFINITION 1.12.– *A Markov chain* $\xi_k^{E_0} = \xi_{\nu_k}$ *with transition probabilities matrix* P^{E_0} *will be called the lumped Markov chain, and it is obtained from the chain* ξ_n *by state lumping from* E *to* E_0.

Let $\{\xi_n, n \geq 0\}$ be a homogeneous Markov chain with the state space $E = \{1, 2, \cdots\}$, which satisfies the condition:

C1. *A Markov chain* $\{\xi_k\}$ *is irreducible (Meyn and Tweedie 1993) and it has the stationary distribution* $\boldsymbol{\rho} = \{\rho_i, i \in E\}$. *In addition,*

$$\varepsilon_n = \sum_{i>n} \rho_i > 0.$$

Denote by e_n the new state obtained after merging the set of states $E^n = \{n+1, n+2, \ldots\}$ (Korolyuk *et al.* 1979a,b), i.e. the transition probabilities of the lumped Markov chain $\{\overline{\xi}_m, m \in \mathbb{N}\}$ on the state space $E_n \cup \{e_n\}$, where $E_n = \{1, 2, \ldots, n\}$, are as follows:

$$\overline{p}_{ij} = \begin{cases} p_{ij}, & i, j \in E_n, \\ \sum_{k \in E^n} p_{ik}, & i \in E_n,\ j = e_n, \\ \frac{1}{\varepsilon_n} \sum_{k \in E^n} \rho_k p_{kj}, & i = e_n,\ j \in E_n, \\ 1 - \sum_{l \in E_n} \overline{p}_{e_n l}, & i = e_n,\ j = e_n. \end{cases}$$

It is easily seen that the stationary distribution $\boldsymbol{\rho}^{(n)}$ of the chain $\{\overline{\xi}_m\}$ is of the form $\boldsymbol{\rho}^{(n)} = (\rho_1, \rho_1, \cdots, \rho_n, \varepsilon_n)$.

Consider a Markov process $\{\xi_\varepsilon(t), t \geq 0\}$, on the state space (X, Σ), where Σ is a σ-algebra of subsets of X. We introduce the following finite partition of the state space $X = \cup_{i=1}^r X_i,\ X_i \in \Sigma$, where $X_i \cap X_j = \emptyset,\ i \neq j$. The process $\xi_\varepsilon(t)$ depends on $\varepsilon > 0$ (where ε is the small series parameter) and its infinitesimal operator A_ε is of the form

$$A_\varepsilon = A + \varepsilon A_1,$$

where A is the infinitesimal operator of the support Markov process $\xi_0(t),\ t \geq 0$, which is obtained from $\xi_\varepsilon(t)$ by letting $\varepsilon = 0$. We assume that $\xi_0(t)$ is uniformly ergodic in X with the transition probabilities $P(x, B)$, $x \in X$, $B \in \Sigma$, which satisfy the following condition:

$$P(x, X_i) = \mathbb{I}_i(x) = \begin{cases} 1, & x \in X_i, \\ 0, & x \notin X_i. \end{cases}$$

We also assume that for all $i = 1, 2, \ldots, r$, there is not a proper subset $B_i \subset X_i$ such that $P(x, B_i) = 1\ \forall x \in B_i$ and $(x, X_i \backslash B_i) = 1\ \forall x \in X_i \backslash B_i$.

Denote by $\pi_i(B)$, $B \in \Sigma$, $\pi_i(X_i) = 1$, $1 \leq i \leq r$ the stationary distribution of the support process $\xi_0(t)$. As mentioned above, the infinitesimal operator A of $\xi_0(t)$ is reducible-invertible (Korolyuk and Turbin 1993).

The kernel of A, say $ker\,(A)$, has dimension r and its proper projector Π is as follows:

$$\Pi f\,(x) = \sum_{i=1}^{r} \int_{X_i} f\,(y)\, \pi_i(dy) \mathbb{I}_i\,(x)\,.$$

The contracted operator $\hat{A}_1$ is defined in the following form:

$$\Pi A_1 \Pi = \hat{A}_1 \Pi.$$

It was shown in Korolyuk and Turbin (1976) that the operator $\hat{A}_1$ can be represented in the matrix form $\hat{A}_1 = [a_{ij},\; i,j \in \{1,2,\ldots,r\}]$, and it is the infinitesimal operator of the new merged or lumped Markov process $\widehat{\xi}(t)$, $t \geq 0$, with the state space $\hat{X} = \{1,2,\ldots,r\}$.

Now, we define the following merging function $k\,(x) = i$ for $x \in X_i$.

It was proved in Korolyuk and Turbin (1976) and Korolyuk and Turbin (1982) that $k\,(\xi_\varepsilon\,(t/\varepsilon))$ weakly converges to $\widehat{\xi}\,(t)$ following weak convergence to $\widehat{\xi}\,(t)$ as $\varepsilon \to 0$ for all $t > 0$, i.e. for all $t > 0$:

$$P\left\{k\,(\xi_\varepsilon\,(t/\varepsilon)) < x\right\} \to P\left\{\widehat{\xi}\,(t) < x\right\} \quad \text{as}\; \varepsilon \to 0, \quad x \in \mathbb{R}.$$

1.6. Switched processes in Markov and semi-Markov media

In this section, we introduce the switched process driven by Markov or semi-Markov process, and it represents the stochastic media, where the switched process evolves. The semigroup of operators associated with a switched process is defined and its infinitesimal operator is represented. In addition, we consider the superposition of independent semi-Markov processes.

Denote by $\{\xi\,(t)\,,\, t \geq 0\}$, a semi-Markov process in a standard state or phase space (X,Σ) given by the semi-Markov kernel

$$Q\,(x, B, t) = P\,(x, B)\, G_x(t).$$

The evolutionary switched process $V(t)$ in a semi-Markov media $\xi\,(t)$ is defined as a solution of the evolution equation (Korolyuk 1993; Korolyuk and Swishchuk 1995b; Anisimov 1977):

$$\frac{dV(t)}{dt} = C\,(V(t), \xi(t))\,, \qquad V(0) = v \tag{1.5}$$

where $C(v,x)\;:\;\mathbb{R}{\times}X \to \mathbb{R}$ satisfies the unique valued solvability condition of equation [1.5]. A sufficient condition for this is that the function $C(v,x)$ satisfies a Lipschitz condition with respect to the variable v uniformly in $x \in X$.

Such a stochastic process with reflecting boundaries is a good model for a multiphase supplying system with feedback, which will be studied in Volume 1, Chapter 3. For instance, the estimation of effectiveness of a supplying system with feedback is reduced to the calculations of the stationary distribution of a switched process that models the system.

Denote by τ_k, $k = 1, 2, \ldots$, the renewal instants of a semi-Markov process $\xi(t)$. Then, equation [1.5] can be solved sequentially in each of the segments $[\tau_n, \tau_{n+1})$, where it is deterministic, i.e.

$$\frac{dV(t;\xi_n)}{dt} = C(V(t;\xi_n), \xi_n), \quad t \in [\tau_n, \tau_{n+1}),$$

with boundary condition $\xi_n = \xi(\tau_n)$, where $V(\tau_n;\xi_n) = V(\tau_{n-1} - 0;\xi_{n-1})$, $n \geq 2$.

The following recurrence relation represents a solution of equation [1.5]

$$V(t) = V(t;\xi_n), \quad t \in [\tau_n, \tau_{n+1}), n \geq 0.$$

Since a solution $V(t)$ of equation [1.5] depends on $x = \xi(t)$ and v, i.e. $V(t) = V(t, x, v)$.

On the space of boundary functions $C_b(R^n)$, we introduce a family of operators T_t, which is defined as follows:

$$T_t(x) f(v) = f(V(t)) = f(V(t, x, v)). \qquad [1.6]$$

LEMMA 1.6.– (Korolyuk 1993; Korolyuk and Swishchuk 1995b) The family of operators $T_t(x)$, $t \geq 0$, introduced in equation [1.6], satisfies the semigroup property, namely for any t, $t' \geq 0$,

$$T_t(x) T_{t'}(x) = T_{t+t'}(x).$$

We should mention that the process $V(t)$ is not Markov even if $\xi(t)$ is Markov. In the case of the Markov switching process $\xi(t)$, it is found that the couple of processes $(V(t), \xi(t))$ is Markov and its infinitesimal operator is defined by the following lemma:

LEMMA 1.7.– (Korolyuk 1993; Korolyuk and Korolyuk 1999) The infinitesimal operator $G(x)$ of the couple of processes $(V(t), \xi(t))$ is of the following form:

$$Gf(v,x) = Qf(v,x) + C(v,x)\frac{\partial}{\partial v} f(v,x),$$

where $f(v,x)$ is a bounded function differentiable with respect to v, and $Qf(v,x) = \int_E Q(x,dy) f(v,y) - \lambda_x f(v,x)$ is the infinitesimal operator of the Markov process $\xi(t)$.

As we mentioned above, the process $V(t)$ with delaying barriers models a multiphase system with feedback. In order to investigate the efficiency of such systems, it is enough to calculate the stationary distribution of the couple of processes $(V(t), \xi(t))$. For such a purpose, we obtain the operator G^* adjoint to the infinitesimal operator G of $(V(t), \xi(t))$ with the same boundary conditions as the process $V(t)$, and we should find a solution of the following equation:

$$G^* \rho = 0.$$

The normalized solution ρ of this equation is the stationary distribution of $(V(t), \xi(t))$.

Another important result used in the study of multiphase systems is the superposition of independent semi-Markov processes. Assuming the case when the semi-Markov process is completely described by its corresponding MRP, we define a superposition of semi-Markov processes by using the superposition of MRPs.

Consider N independent MRPs $\left\{\xi_n^{(i)}, \tau_n^{(i)}\right\}$, where $\tau_n^{(i)}$ *are renewal instants* of the ith MRP. Let us introduce an auxiliary process as follows:

$$\gamma^{(i)}(t) = \inf\left\{s > t : \xi_n^{(i)}(s) \neq \xi_n^{(i)}(t)\right\}.$$

Consider the processes

$$\gamma(t) = \min\left\{\gamma^{(1)}(t), \gamma^{(2)}(t), \cdots, \gamma^{(N)}(t)\right\};$$

$$\zeta^{(i)}(t) = \gamma^{(i)}(t) - \gamma(t).$$

Then the renewal instants $\{\tau_n, n \geq 1\}$ of the superposition of MRPs are defined by the following relations:

$$\gamma(\tau_n - 0) = 0, \ n \geq 1.$$

Define

$$\zeta_n^{(i)} = \gamma^{(i)}(\tau_{n-1}) - \gamma(\tau_{n-1}), \ n \geq 1.$$

DEFINITION 1.13.– (Korolyuk and Turbin 1982) *The Markov renewal process* $\{\xi_n, \zeta_n\}$ *with components* $\xi_n = \left\{\xi_n^{(1)}, \xi_n^{(2)}, \cdots, \xi_n^{(N)}\right\}$ *and* $\zeta_n = \left\{\zeta_n^{(1)}, \zeta_n^{(2)}, \cdots, \zeta_n^{(N)}\right\}$ *is called the superposition of Markov renewal processes* $\left\{\xi_n^{(i)}, \tau_n^{(i)}\right\}$, $i = 1, 2, \cdots, N$.

The superposition of processes models a multiphase system with feedback, which consists of several consecutive aggregates such that in between any two successive aggregates there is a reservoir. This type of system will be studied in Volume 1, Chapter 3.

2

Homogeneous Random Evolutions (HRE) and their Applications

In mathematical language, a random evolution (RE) is a solution of a stochastic operator integral equation in a Banach space. The operator coefficients of such equations depend on random parameters. The RE, in physical language, is a model for a dynamical system whose state of evolution is subject to random variations. Such systems arise in many branches of science, e.g. random Hamiltonian and Shrödinger's equations with random potential in quantum mechanics, Maxwell's equation with a random reflective index in electrodynamics, transport equation and storage equation. In addition, there are many applications of REs in financial and insurance mathematics (Swishchuk 2000). One of the recent applications of RE is associated with geometric Markov renewal processes (GMRPs), which are regime-switching models for a stock price in financial mathematics. These models will be studied extensively in the following chapters. Another interesting application of RE is a semi-Markov risk process in insurance mathematics (Swishchuk 2000). The REs are also examples of more general mathematical objects such as multiplicative operator functionals (MOFs) (Pinsky 1991; Swishchuk 1997), which are random dynamical systems in Banach space. REs can be described by two mathematical objects: (1) an operator dynamical system $V(t)$ and (2) a random process x_t. Depending on the structure of $V(t)$ and on properties of the stochastic process x_t, we have different kinds of REs: continuous, discrete, Markov, semi-Markov, etc. In this chapter, we deal with various issues related to REs, including the martingale property, the asymptotical behaviors of REs, averaging, merging, diffusion approximation, normal deviations, averaging and diffusion approximation in reducible phase for x_t, and rate of convergence for limit theorems for REs. Note that inhomogeneous REs and their applications were considered in Swishchuk (2020).

2.1. Homogeneous random evolutions (HRE)

2.1.1. *Definition and classification of HRE*

Let $(\Omega, \mathcal{F}, \mathcal{F}_t, \mathbb{P})$ be a probability space, $t \in \mathbb{R}_+ := [0, +\infty]$, let (X, Ξ) be a measurable phase space and let $(B, \mathcal{B}, \| \cdot \|)$ be a separable Banach space.

Let us consider a Markov renewal process $(x_n, \theta_n; n \geq 0)$, $x_n \in X$, $\theta_n \in \mathbb{R}_+$, $n \geq 0$, with stochastic kernel

$$Q(x, A, t) := P(x, A)G_x(t), \quad P(x, A) := P(x_{n+1} \in A | \, x_n = x),$$

$$G_x(t) := P(\theta_{n+1} | \, x_n = x), \qquad [2.1]$$

$x \in X, a \in \Xi, t \in \mathbb{R}_+$. Process $x_t := x_{\nu(t)}$ is called a semi-Markov process, where $\nu(t) := \max\{n : \tau_n \leq t\}$, $\tau_n := \sum_{k=0}^{n} \theta_k$, $x_n = x_{\tau_n}$, $P\{\nu(t) < +\infty, \forall t \in \mathbb{R}_+\} = 1$. Note that if $G_x(t) = 1 - e^{-\lambda(x)t}$, where $\lambda(x)$ is a measurable and bounded function on X, then x_t is called a jump Markov process.

Let $\{\Gamma(x); x \in X\}$ be a family of operators on the dense subspace $B_0 \in B$, which is a common domain for $\Gamma(x)$, independent of x, non-commuting and unbounded in general, such that map $\Gamma(x)f : X \to B$ is strongly $\Xi/\mathcal{B}$-measurable for all $f \in B$, $\forall t \in \mathbb{R}_+$; also, let $\{\mathcal{D}(x, y); x, y \in X\}$ be a family of bounded linear operators on B, such that map $\mathcal{D}(x, y)f : X \times X \to B$ is $\Xi \times \Xi/\mathcal{B}$-measurable, $\forall f \in B$.

An **RE** is defined by the solution of a stochastic operator integral equation in separable Banach space B:

$$V(t)f = f + \int_0^t V(s)\Gamma(x_s)f ds + \sum_{k=1}^{\nu(t)} V(\tau_k^-)[\mathcal{D}(x_{k-1}, x_k) - I]f, \qquad [2.2]$$

where I is an identity operator on B, $\tau_k^- := \tau_k - 0, f \in B$.

If x_t in [2.1] is a Markov or semi-Markov process, then the RE in [2.2] is called a **Markov or semi-Markov RE**, respectively.

If $\mathcal{D}(x, y) \equiv I, \forall \, x, y \in X$, then $V(t)$ in [2.2] is called **a continuous RE**.

If $\Gamma(x) \equiv 0, \forall \, x \in X$, is a zero operator on B, then $V(t)$ in [2.2] is called **a jump RE**.

RE $V_n := V(\tau_n)$ is called **a discrete RE**.

Operators $\Gamma(x)$, $x \in X$, describe a continuous component $V^c(t)$ of RE $V(t)$ in [2.2], and operators $\mathcal{D}(\S, \dagger)$ describe a jump component $V^d(t)$ of RE $V^d(t)$ in [2.2].

In this manner, a RE is described by two objects: (1) an operator dynamical system $V(t)$ and (2) a random process x_t.

Note that the operator $V(t)$ turns out to be (Korolyuk and Swishchuk 1995a,b)

$$V(t) = \Gamma_{x_t}(t - \tau_{\nu(t)}) \prod_{k=1}^{\nu(t)} \mathcal{D}(x_{k-1}, x_k)\Gamma_{x_{k-1}}(\theta_k), \qquad [2.3]$$

where $\Gamma_x(t)$ are the semigroups of operators of t generated by the operators $\Gamma(x)$, $\forall x \in X$. We also note that RE in [2.2] is usually called **a discontinuous RE**. Under the above-introduced conditions, the solution $V(t)$ of equation [2.2] is unique and can be represented by the product [2.3], which can be proved by a constructive method (Korolyuk and Swishchuk 1995b).

REMARK 2.1.– *From the definition of REs given above, it follows that there are other examples of MOFs, as they satisfy all the conditions for MOFs.*

2.1.2. *Some examples of HRE*

The connection of RE with applied problems can be explained by the generality of definition [2.2] of RE. It includes any homogeneous linear evolutionary system. For instance, if

$$\Gamma(x) := v(x)\frac{d}{dz}, \quad \mathcal{D}(x,y) \equiv I, \quad B = C^1(R),$$

then equation [2.2] is a transport equation that describes the motion of a particle with random velocity $v(x_t)$. Thus, various interpretations of operators $\Gamma(x)$ and $\mathcal{D}(x,y)$ give us many realizations of RE. To illustrate the flexibility and richness of this approach, we provide some examples.

EXAMPLE 2.1. (Impulse traffic process).– Let $B = C(R)$ and operators $\Gamma(x)$ and $\mathcal{D}(x,y)$ are defined as

$$\Gamma(x)f(z) := v(z,x)\frac{d}{dz}f(z), \quad \mathcal{D}(x,y)f(z) := f(z + a(x,y)), \qquad [2.4]$$

where the functions $v(z,x)$ and $a(x,y)$ are continuous and bounded on $R \times X$ and $X \times X$, respectively, $\forall z \in R, \forall x,y \in X$, $f(z) \in C^1(R) := B_0$. Then, equation [2.2] takes the form

$$f(z_t) = f(z) + \int_0^t v(z_s, x_s)\frac{d}{dz}f(z_s)\,ds$$

$$+\sum_{k=1}^{\nu(t)}[f(z_{\tau_k^-}+a(x_{k-1},x_k))-f(z_{\tau_k^-})], \qquad [2.5]$$

and RE $V(t)$ is defined by the relation

$$V(t)f(z)=f(z_t),$$

$z_0 = z$. Equation [2.5] is a functional one for **impulse traffic process** z_t, which satisfies the equation

$$z_t=z+\int_0^t v(z_s,x_s)ds+\sum_{k=1}^{\nu(t)}a(x_{k-1},x_k). \qquad [2.6]$$

Note that the impulse traffic process z_t in [2.6] is a realization of a discontinuous RE.

EXAMPLE 2.2. (Summation on a Markov chain).– Let us assume $v(z,x)\equiv 0, \forall\, z\in R$, $\forall\, x\in X$, in [2.6]. Then, the process

$$z_t=z+\sum_{k=1}^{\nu(t)}a(x_{k-1},x_k) \qquad [2.7]$$

is a summation on a Markov chain $(x_n; n\geq 0)$ and it is a realization of a jump RE. Let $z_n := z_{\tau_n}$ in [2.7], then the discrete process

$$z_n=z+\sum_{k=1}^{n}a(x_{k-1},x_k)$$

is a realization of a discrete RE.

EXAMPLE 2.3. (Diffusion process in random media).– Let $B=C(\mathbb{R})$, $B_0=C^2(\mathbb{R})$, $P_x(t,z,A)$ be a Markov continuous distribution function with respect to the diffusion process $\xi(t)$, i.e. the solution of the stochastic differential equation in $\mathbb{R}$ with semi-Markov switching:

$$d\xi(t)=\mu(\xi(t),x_t)dt+\sigma(\xi(t),x_t)dw_t, \quad \xi(0)=z, \qquad [2.8]$$

where x_t is a semi-Markov process independent of a standard Wiener process w_t, coefficients $\mu(z,x)$ and $\sigma(z,x)$ are bounded and continuous functions on $\mathbb{R}\times X$. Let us define the following contraction semigroups of operators on B:

$$\Gamma_x(t)f(z):=\int_{\mathbb{R}}P_x(t,z,dy)f(y), \quad f(y)\in B,\ x\in X. \qquad [2.9]$$

Their infinitesimal operators $\Gamma(x)$ are of the following kind:

$$\Gamma(x)f(z) = \mu(z,x)\frac{d}{dz}f(z) + \frac{1}{2}\sigma^2(z,x)\frac{d^2}{dz^2}f(z), \quad f(z) \in B_0.$$

Then the process $\xi(t)$ is a continuous one. This is the main reason why the operators $\mathcal{D}(\S,\dagger) \equiv \mathcal{I}, \forall x, y \in X$, are identity operators. Hence, equation [2.2] takes the form

$$f(\xi(t)) = f(z) + \int_0^t \left[\mu(\xi(s),x_s)\frac{d}{dz} + \frac{\sigma^2(\xi(s),x_s)}{2}\frac{d^2}{dz^2}\right] f(\xi(s))ds, \quad [2.10]$$

and the RE $V(t)$ is defined by the relation

$$V(t)f(z) = \mathbf{E}[f(\xi(t))|\, x_s;\, 0 \le s \le t;\, \xi(0) = z].$$

Equation [2.10] is a functional for the diffusion process $\xi(t)$ in [2.8] in a semi-Markov random media x_t. Note that diffusion process $\xi(t)$ in [2.8] is a realization of a continuous RE.

EXAMPLE 2.4. (The geometric Markov renewal process (GMRP)).– (Swishchuk and Islam 2010). Let $(x_n, \theta_n)_{n\in\mathbb{Z}_+}$ be a Markov renewal process on the phase space $X \times \mathbb{R}_+$ with the semi-Markov kernel $Q(x, A, t)$ and $x(t) := x_{\nu(t)}$ be a semi-Markov process. Let $\rho(x)$ be a bounded continuous function on X such that $\rho(x) > -1$. We define a stochastic functional S_t with a Markov renewal process $(x_n;\, \theta_n)_{n\in\mathbb{Z}_+}$ as follows:

$$S_t := S_0 \prod_{k=0}^{\nu(t)} (1 + \rho(x_k)), \quad [2.11]$$

where $S_0 > 0$ is the initial value of S_t. We call the process $(S_t)_{t\in\mathbb{R}_+}$ in [2.11] a GMRP. This process $(S_t)_{t\in\mathbb{R}_+}$ is referred to as GMRP by analogy with the geometric compound Poisson process

$$S_t = S_0 \prod_{k=1}^{N(t)} (1 + Y_k), \quad [2.12]$$

where $S_0 > 0$, $N(t)$ is a standard Poisson process, and $(Y_k)_{k\in\mathbb{Z}_+}$ are independent and identically distributed (i.i.d.) random variables. The GMRP is a trading model in many financial applications as a pure jump model (see Swishchuk and Islam (2010) and Part 2 of Volume 2 of this book).

Let $B : C_0(\mathbb{R}_+)$ be a space of continuous functions on $\mathbb{R}_+$, vanishing at infinity, and let us define a family of bounded contracting operators $D(x)$ on $C_0(\mathbb{R}_+)$:

$$D(x)f(s) := f(s(1+\rho(x)), \quad x \in X, \ s \in \mathbb{R}_+. \qquad [2.13]$$

With these contraction operators $D(x)$, we define the following jump semi-Markov random evolution (JSMRE) $V(t)$ of GMRP in [2.11]

$$V(t) = \prod_{k=0}^{\nu(t)} D(x_k) := D(x_{\nu(t)}) \circ D(x_{\nu(t)-1}) \circ \ldots \circ D(x_1) \circ D(x_0). \quad [2.14]$$

Using [2.13], we obtain from [2.14]

$$V(t)f(s) = \prod_{k=0}^{\nu(t)} D(x_k)f(s) = f(s \prod_{k=0}^{\nu(t)} (1+\rho(x_k)) = f(S_t), \qquad [2.15]$$

where S_t is defined in [2.11] and $S_0 = s$.

2.1.3. *Martingale characterization of HRE*

One of the main approaches to the study of RE are the martingale methods. The main idea is that process

$$M_n := V_n - I - \sum_{k=0}^{n-1} \mathbf{E}[V_{k+1} - V_k | \mathcal{F}_k], \quad V_0 = I, \qquad [2.16]$$

is an $\mathcal{F}_n$-martingale in B, where

$$\mathcal{F}_n := \sigma\{x_k, \tau_k; \ 0 \le k \le n\}, \quad V_n := V(\tau_n),$$

$\mathbf{E}$ is the expectation operator for probability measure $\mathbb{P}$. The representation of the martingale M_n (see [2.4]) is in the form of a martingale difference such as the following one:

$$M_n = \sum_{k=0}^{n-1} [V_{k+1} - \mathbf{E}(V_{k+1} | \mathcal{F}_k)] \qquad [2.17]$$

which gives us the possibility to calculate the weak quadratic variation

$$< l(M_n f) > := \sum_{k=0}^{n-1} \mathbf{E}[l^2((V_{k+1} - V_k)f) | \mathcal{F}_k], \qquad [2.18]$$

where $l \in B^*$, and B^* is a dual space to B, dividing points of B.

The martingale method for obtaining limit theorems for the sequence of RE is founded on the solution of the following problems: (1) weak compactness of the family of measures generated by the sequences of RE; (2) any limiting point of this family of measures is the solution of the martingale problem; (3) the solution of the martingale problem is unique. The conditions (1) and (2) guarantee the existence of weakly converging subsequence, and condition (3) gives the uniqueness of the weak limit. It follows from (1) to (3) that a solution of RE converges weakly to the unique solution of the corresponding martingale problem. The weak convergence of RE is obtained as a series scheme from the criterion of weak compactness of processes with values in a separable Banach space (Korolyuk and Swishchuk 1995b). The limit solution of a RE is obtained from the solution of a corresponding martingale problem in the form of some integral operator equations in Banach space B. We also use the representation

$$V_{k+1} - V_k = [\Gamma_{x_k}(\theta_{k+1})\mathcal{D}(x_k, x_{k+1}) - I]V_k, \quad V_k := V(\tau_k), \qquad [2.19]$$

and the following expression for semigroups of operators $\Gamma_x(t)$ (Korolyuk and Swishchuk 1995b):

$$\Gamma_x(t)f = f + \sum_{k=1}^{n-1} \frac{t^k}{k!}\Gamma^k(x)f + \frac{1}{(n-1)}\int_0^t (t-s)^n \Gamma_x(s)\Gamma^n(x)f ds, \quad [2.20]$$

$\forall x \in X, \forall f \in \cap_{x\in X} Dom(\Gamma^n(x))$. Taking into account [2.4]–[2.8], we obtain the limit theorems for RE. In the previous section, we considered the evolution equation associated with REs by using the jump structure of the semi-Markov process or jump Markov process. In order to deal with more general driving processes and to consider other applications, it is useful to reformulate the treatment of RE in terms of a **martingale problem**. It has been shown by Stroock and Varadhan that the entire theory of multidimensional diffusion processes (and many other continuous-parameter Markov processes) can be formulated by following this approach. Suppose that we have an evolution equation of the form:

$$\frac{df}{dt} = Gf. \qquad [2.21]$$

The **martingale problem** is to find a Markov process $x(t)$, $t \geq 0$, and a RE $V(t)$ so that for all smooth functions

$$V(t)f(x(t)) - \int_0^t V(s)Gf(x(s))ds \quad \text{is a martingale.} \qquad [2.22]$$

This gives the required solution immediately. Indeed, the operator

$$f \to T(t)f := \mathbf{E}_x[V(t)f(x(t))]$$

defines a semigroup of operators on the Banach space B, whose infinitesimal generator can be computed by taking the expectation:

$$\mathbf{E}_x[V(t)f(x(t))] - f(x) = \mathbf{E}_x\left[\int_0^t V(s)Gf(x(s))ds\right],$$

and

$$\lim_{t\to 0} t^{-1}[\mathbf{E}_x[V(t)f(x(t))] - f(x)] = \lim_{t\to 0} t^{-1}\mathbf{E}_x\left[\int_0^t V(s)Gf(x(s))ds\right] = Gf(x).$$

REMARK 2.2.– *In case $V(t) \equiv I$-identity operator, the above reduces to the usual martingale problem for Markov process (Dynkin 1991).*

REMARK 2.3.– *In case $B = \mathbb{R}$, the problem reduces to the determination of a real-valued multiplicative functional, which is related to a Feynman–Kac type formula. In the case of the one-dimensional Wiener process, a wide class of multiplicative functionals is provided by*

$$V(t) = \exp\left\{\int_0^t a(x(s))ds + \int_0^t b(x(s))dw(s)\right\},$$

where $w(t)$ is a standard Wiener process.

Now, let us illustrate the martingale problem for a discontinuous RE over a jump Markov process, diffusion process, etc.

Martingale problem for discontinuous RE over a jump Markov process

Let $x(t)$, $t \geq 0$, be a conservative regular jump Markov process on a measurable state space (X, Ξ) with rate function $\lambda(x) > 0$ and a family of probability measures $P(x, dy)$. In addition, let $V(t)$ be a discontinuous RE in [2.2]. For any Borel function f, we have the sum

$$f(x(t)) = f(x(0)) + \sum_{0\leq s\leq t} [f(x(s+0)) - f(x(s-0))]. \qquad [2.23]$$

From this, we see that the product $V(t)f(x(t))$ satisfies the differential equation:

$$\frac{dV(t)f(x(t))}{dt} = V(t)\Gamma(x(t))f(x(t)), \quad \text{if} \quad \tau_k < t < \tau_{k+1},$$

and the jump across $t = \tau_k$ is evaluated as

$$V(t)f(x(t))|_{\tau_k^-}^{\tau_k^+} = V(\tau_k^-)\mathcal{D}(x(\tau_k^-), x(\tau_k^+))f(x(\tau_k+0)) - f(x(\tau_k-0))$$

leading to the equation:

$$V(t)f(x(t)) = f(x) + \int_0^t V(s)\Gamma(x(s))f(x(s))ds$$
$$+ \sum_{0\leq\tau_k\leq t} V(\tau_k^-)[\mathcal{D}(x(\tau_k^-), x(\tau_k^+))f(x(\tau_k^+)) - f(x(\tau_k^-))], \qquad [2.24]$$

$x(0) = x$, $\tau_k^{\pm} := \tau_k \pm 0$. To state this in the appropriate form of the martingale problem, we use the following identity from the theory of Markov processes: for any positive Borel-measurable function $\phi(\cdot,\cdot)$, we have

$$\mathbf{E}_x\left[\sum_{0\leq\tau_k\leq t} \phi(x(\tau_k^-), x(\tau_k^+))\right]$$
$$= \mathbf{E}_x\left[\int_0^t \lambda(x(s))\int_X \phi(x(s), y)P(x(s), dy)ds\right]. \qquad [2.25]$$

Note that the difference

$$\sum_{0\leq\tau_k\leq t} \phi(x(\tau_k^-), x(\tau_k^+)) - \int_0^t \lambda(x(s))(P\phi)(x(s))ds$$

is a martingale, where P is an operator generated by $P(x, A)$, $x \in X$, $A \in \Xi$. Applying this to the above computations, we see that

$$V(t)f(x(t)) = f(x) + \int_0^t V(s)Gf(x(s))ds + Z(t), \qquad [2.26]$$

where $Z(t), t \geq 0$, is a martingale and

$$Gf(x) = \Gamma(x)f + \lambda(x)\int_X [\mathcal{D}(x, y)f(y) - f(x)]P(x, dy).$$

Martingale problem for discontinuous RE over semi-Markov process

It is well known that process $(x(t), \gamma(t))$ (with $\gamma(t) := t - \tau_{\nu(t)}$ and $x(t)$ as semi-Markov process) is a Markov process in $X \times \mathbb{R}_+$ with infinitesimal operator

$$\hat{Q} := \frac{d}{dt} + \frac{g_x(t)}{\bar{G}_x(t)}[P - I],$$

where $g_x(t) := dG_x(t)/dt$, $\bar{G}_x(t) := 1 - G_x(t)$, P is an operator generated by $P(x, A)$, $x \in X$, $A \in \Xi$, $P(x, A)$ and $G_x(t)$ are defined in [2.1]. In the Markov case, $G_x(t) = 1 - \exp\{-\lambda(x)t\}$, $g_x(t) = \lambda(x)\exp\{-\lambda(x)t\}$, $\hat{G}_x(t) = \exp\{-\lambda(x)t\}$, and $g_x(t)/\hat{G}_x(t) = \lambda(x)$, $\forall x \in X$. Hence, $\hat{Q} = \lambda(x)[P - I]$ is an infinitesimal operator of a jump Markov process $x(t)$ in X. Using the reasoning [2.23]–[2.26] of the previous example, for Markov process $y(t) := (x(t), \gamma(t))$ in $X \times \mathbb{R}_+$ we obtain that the solution for the martingale problem is the operator

$$Gf(x,t) = \frac{d}{dt}f(x,t) + \Gamma(x)f(x,t) + \frac{g_x(t)}{\hat{G}_x(t)}\int_X [\mathcal{D}(\S,\dagger)\{(\dagger,\sqcup) - \{(\S,\prime)]\mathcal{P}(\S,\lceil\dagger),$$

and the process $y(t)$.

Martingale problem for RE over Wiener process

Let $w(t), t \geq 0$, be the Wiener process in $\mathbb{R}^d$ and consider the linear stochastic equation:

$$V(t) = I + \int_0^t V(s)\Gamma_0(w(s))ds + \sum_{j=1}^{d}\int_0^t V(s)\Gamma_j(w(s))dw_j(s),$$

where the final term is a stochastic integral of the Itô class and $\Gamma_0, \ldots, \Gamma_d$ are bounded operators on a Banach space B. If f is any C^2 function, then the formula of Itô gives

$$f(w(t)) = f(w(0)) + \frac{1}{2}\int_0^t \Delta f(w(s))ds + \sum_{j=1}^{d}\int_0^t \frac{\partial f}{\partial w_j}(w(s))dw_j(s).$$

Using the stochastic product rule

$$d(Mf) = Mdf + (dM)f + (dM)df \qquad [2.27]$$

and rearranging terms, we have

$$V(t)f(w(t)) = f(w(0)) + \int_0^t V(s)\left(\frac{1}{2}\Delta f + \sum_{j=1}^{d}\Gamma_j\frac{\partial f}{\partial w_j} + \Gamma_0 f\right)w(s)ds + Z(t),$$

where

$$Z(t) := \sum_{j=1}^{d}\int_0^t V(s)\left(\frac{\partial f}{\partial w_j}(w(s)) + \Gamma_j(w(s))f(w(s))\right)dw_j(s),$$

which is a martingale. Therefore, we have obtained the solution of the martingale problem, with the infinitesimal generator

$$Gf = \frac{1}{2}\Delta f(w) + \sum_{j=1}^{d}\Gamma_j(w)\frac{\partial f}{\partial w_j}(w) + \Gamma_0(w)f(w).$$

This corresponds to the stochastic solution of the parabolic system

$$\frac{\partial u}{\partial t} = Gu.$$

Martingale problem for RE over diffusion process

Let $\xi(t), t \geq 0$, be the diffusion process in $\mathbb{R}$:

$$d\xi(t) = a(\xi(t))dt + \sigma(\xi(t))dw(t)$$

and consider the linear stochastic equation:

$$V(t) = I + \int_0^t V(s)\Gamma_0(\xi(s))ds + \int_0^t V(s)\Gamma_1(\xi(s))d\xi(s),$$

with the bounded operators Γ_0 and Γ_1 on B. If f is any C^2 function, the formula of Itô gives

$$f(\xi(t)) = f(\xi(0)) + \int_0^t \left[a(\xi(s))\frac{df(\xi(s))}{d\xi} + \frac{1}{2}\sigma^2(\xi(s))\frac{d^2 f(\xi(s))}{d\xi^2}\right] ds + \int_0^t \frac{\partial f(\xi(s))}{\partial \xi}\sigma(\xi(s))dw(s).$$

Using the stochastic product rule [2.27], we have:

$$V(t)f(\xi(t)) = f(\xi(0)) + \int_0^t V(s)\left(a\frac{df}{d\xi} + 2^{-1}\sigma^2\frac{d^2 f}{d\xi^2} + \Gamma_1\frac{df}{d\xi} + \Gamma_0 f\right)(\xi(s))ds + Z(t),$$

where

$$Z(t) := \int_0^t V(s)\left(\sigma\frac{df}{d\xi} + \Gamma_1 f\right)(\xi(s))dw(s),$$

which is a martingale. Therefore, we have obtained the solution of the martingale problem with the operator

$$Gf = a\frac{df}{d\xi} + \frac{1}{2}\sigma^2\frac{d^2 f}{d\xi^2} + \Gamma_1\frac{df}{d\xi} + \Gamma_0 f.$$

Other solutions of martingale problems for RE will be obtained in the limit theorems for RE.

2.1.4. *Analogue of Dynkin's formula for HRE*

Let $x(t), t \geq 0$, be a strongly measurable strong Markov process, let $V(t)$ be an MOF of $x(t)$ (Pinsky 1991; Swishchuk 1997), let A be the infinitesimal operator of semigroup

$$(T(t)f)(x) := \mathbf{E}_x[V(t)f(x(t))], \qquad [2.28]$$

and let τ be a stopping time for $x(t)$. It is known (Swishchuk 1997) that if $Ah = g$ and $\mathbf{E}_x[\tau] < +\infty$, then

$$\mathbf{E}_x[V(\tau)h(x(\tau))] - h(x) = \mathbf{E}_x \int_0^\tau V(t)Ah(x(t))dt. \qquad [2.29]$$

Formula [2.28] is an analogue of Dynkin's formula for MOF (Swishchuk 1997). In fact, if we set $V(t) \equiv I$-identity operator, then from [2.29] we obtain

$$\mathbf{E}_x[h(x(\tau))] - h(x) = \mathbf{E}_x \left[\int_0^\tau Qh(x(t))dt \right], \qquad [2.30]$$

where Q is an infinitesimal operator of $x(t)$ (see [2.28]). Formula [2.30] is the well-known Dynkin's formula. Let $x(t)$, $t \geq 0$, be a continuous Markov process on (X, Ξ) and $V(t)$ be a continuous RE, then

$$\frac{dV(t)}{dt} = V(t)\Gamma(x(t)), \quad V(0) = I. \qquad [2.31]$$

It should be noted that the function $u(t,x) := \mathbf{E}_x[V(t)f(x(t))]$ satisfies the following equation (Swishchuk 1997):

$$\frac{du(t,x)}{dt} = Qu(t,x) + \Gamma(x)u(t,x), \quad u(0,x) = f(x), \qquad [2.32]$$

where Q is an infinitesimal operator of $x(t)$. From [2.29] and [2.32], we obtain the **analogue of Dynkin's formula for continuous Markov RE** $V(t)$ in [2.31]. Namely,

$$\mathbf{E}_x[V(\tau)h(x(\tau))] - h(x) = \mathbf{E}_x \int_0^\tau V(t)[Q + \Gamma(x(t))]h(x(t))dt. \qquad [2.33]$$

Let $x(t), t \geq 0$, be a jump Markov process with infinitesimal operator Q and $V(t)$ be a discontinuous Markov RE in [2.2]. In this case, the function $u(t,x) := \mathbf{E}_x[V(t)f(x(t))]$ satisfies the equation (Swishchuk 1997):

$$\frac{du(t,x)}{dt} = Qu(t,x) + \Gamma(x)u(t,x)$$

$$+\lambda(x) \int_X P(x,dy)[\mathcal{D}(x,y) - I]u(t,y), \qquad [2.34]$$

where $u(0,x) = f(x)$. From [2.29] and [2.34], we obtain the **analogue of Dynkin's formula for discontinuous Markov RE** in [2.2]:

$$\mathbf{E}_x[V(\tau)f(x(\tau))] - f(x) = \mathbf{E}_x\left\{\int_0^\tau V(s)\Big[Q + \Gamma(x(t))\right.$$

$$\left.+\lambda(x)\int_X P(x(t),dy)(\mathcal{D}(x(t),y) - I)\Big] f(x(t))dt\right\}. \qquad [2.35]$$

Let us finally assume that $x(t), t \geq 0$, is a semi-Markov process, and $V(t)$ is a semi-Markov RE in [2.2]. Let us define the process

$$\gamma(t) := t - \tau_{\nu(t)}. \qquad [2.36]$$

Then, the process

$$y(t) := (x(t), \gamma(t)) \qquad [2.37]$$

is a Markov process in $X \times \mathbb{R}_+$ with infinitesimal operator (Korolyuk and Swishchuk 1995b)

$$\hat{Q} := \frac{d}{dt} + \frac{g_x(t)}{\bar{G}_x(t)}[P - I], \qquad [2.38]$$

where $g_x(t) := dG_x(t)/dt$, $\bar{G}_x(t) := 1 - G_x(t)$, P is an operator generated by the kernel $P(x,A)$. Hence, the process $(V(t)f;\ x(t);\ \gamma(t);\ t \geq 0) \equiv (V(t)f;\ y(t);\ t \geq 0)$ in $B \times X \times \mathbb{R}_+$ is a Markov process with infinitesimal operator

$$L(x) := \hat{Q} + \Gamma(x) + \frac{g_x(t)}{\bar{G}_x(t)}\int_X P(x,dy)[\mathcal{D}(x,y) - I], \qquad [2.39]$$

where $\hat{Q}$ is defined in [2.38].

Let $f(x,t)$ be a function on $X \times \mathbb{R}_+$ bounded by x and differentiable by t, and let τ be a stopping time for $y(t) = (x(t),\ \gamma(t))$. Then for semi-Markov RE $V(t)$ in [2.2] we have from [2.29] and [2.36]–[2.39] the following **analogue of Dynkin's formula**:

$$\mathbf{E}_y[V(\tau)f(y(\tau))] - f(y) = \mathbf{E}_y\left\{\int_0^\tau V(s)\Big[\hat{Q} + \Gamma(x(t))\right.$$

$$\left.+\frac{g_x(t)}{\bar{G}_x(t)}\int_X P(x(t),dy)[\mathcal{D}(x(t),y) - I]\Big] f(y(t))dt\right\}, \qquad [2.40]$$

where $y := y(0) = (x,0), f(y) = f(x,0)$.

2.1.5. *Boundary value problems for HRE*

Let $x(t), t \geq 0$, be a continuous Markov process in semicompact state space (X, Ξ). Let $V(t)$ be a continuous Markov RE in [2.31], and let G be an open set satisfying $\forall x \in G$; there exists $U : \mathbf{E}_x[\tau_U] < +\infty,\ U \in \Xi$, and

$$\tau_U := \inf_t \{t : x(t) \notin U\},\ P_x[\tau_G = +\infty] = 0, \forall x \in X. \qquad [2.41]$$

If $f(x)$ is a bounded measurable function on ∂G (boundary of G) and function

$$b(x) := \mathbf{E}_x[V(\tau_G) f(x(\tau_G))] \qquad [2.42]$$

is continuous on X, then function $b(x)$ is the solution of the equation (Swishchuk 1997):

$$Qb(x) + \Gamma(x)b(x) = 0, \quad \forall x \in G, \qquad [2.43]$$

where Q is an infinitesimal operator of $x(t)$. If function

$$H(x) := \mathbf{E}_x \left[\int_0^{\tau_G} V(t) g(x(t)) dt \right] \qquad [2.44]$$

is continuous and bounded, then this function satisfies the following equation (Swishchuk 1997):

$$QH(x) + \Gamma(x)H(x) = -g(x), \quad \forall x \in X. \qquad [2.45]$$

It follows from [2.41] to [2.44] that the boundary value problem

$$QH(x) + \Gamma(x)H(x) = -g(x), \quad H(x)|_{\partial G} = f(x) \qquad [2.46]$$

has the following solution:

$$H(x) = \mathbf{E}_x \left[\int_0^{\tau_G} V(s) g(x(s)) ds \right] + \mathbf{E}_x[V(\tau_G f(x(\tau_G))]. \qquad [2.47]$$

Now, let $x(t), t \geq 0$, be a jump Markov process in (X, Ξ), let $V(t)$ be a discontinuous Markov RE in [2.2] and let conditions [2.41] be satisfied. It follows from [2.44] to [2.47] and from [2.44] that the boundary value problem

$$QH(x) + \Gamma(x)H(x) + \int_X P(x, dy)[\mathcal{D}(x, y) - I]H(y) = -g(x), H(x)|_{\partial G} = f(x)$$

has the solution

$$H(x) = \mathbf{E}_x \left[\int_0^{\tau_G} V(s) g(x(s)) ds \right] + \mathbf{E}_x[V(\tau_G) f(x(\tau_G))].$$

2.2. Limit theorems for HRE

The main approach to the investigation of limit theorems for SMRE is the martingale method.

The martingale method for obtaining the limit theorems (averaging and diffusion approximation) for the sequence of SMRE is bounded on the solution of the following problems:

1) weak compactness of the family of measures generated by the sequence of SMRE;

2) any limiting point of this family of measures is the solution of martingale problem;

3) the solution of martingale problem is unique.

The conditions (1) and (2) guarantee the existence of a weakly converging subsequence, and condition (3) gives the uniqueness of a weak limit.

From (1) to (3), it follows that SMRE converges weakly to the unique solution of martingale problem.

2.2.1. *Weak convergence of HRE*

A **weak convergence of SMRE** in a series scheme can be obtained from the criterion of weak compactness of the process with values in a separable Banach space (Korolyuk and Swishchuk 1995b). The **limit SMRE** that we obtain from the solution of the corresponding martingale problem is a kind of integral operator equation in Banach space B.

The main idea is that process

$$M_n := V_n - I - \sum_{k=0}^{n-1} \mathbf{E}\left[(V_{k+1} - V_k)|\, \mathcal{F}_k\right], \quad V_0 = I, \qquad [2.48]$$

is an $\mathcal{F}_n$–**martingale** in B, where

$$\mathcal{F}_n := \sigma\{x_k, \tau_k;\ 0 \le k \le n\}, \quad V_n := V(\tau_n),$$

E is an expectation by probability $\mathbb{P}$ on a probability space $(\Omega, \mathcal{F}, \mathbb{P})$.

The representation of the martingale M_n in the form of **martingale-differences**

$$M_n = \sum_{k=0}^{n-1} [V_{k+1} - \mathbf{E}(V_{k+1}|\, \mathcal{F}_k)] \qquad [2.49]$$

gives us the possibility to calculate the **weak quadratic variation**:

$$< l(M_n f) >:= \sum_{k=0}^{n-1} \mathbf{E}\left[l^2((V_{k+1} - V_k)f) | \mathcal{F}_k\right], \qquad [2.50]$$

where $l \in B^*$, and B^* is a dual space to B, dividing points of B.

From [2.19], it follows that

$$V_{k+1} - V_k = [\Gamma_{x_k}(\theta_{k+1}) D(x_k, x_{k+1}) - I] \cdot V_k. \qquad [2.51]$$

Note that the following expression for semigroup of operators $\Gamma_x(t)$ is fulfilled:

$$\Gamma_x(t) f = I + \sum_{k=1}^{n-1} \frac{t^k}{k!} \Gamma^k_{(x)} f + \frac{1}{(n-1)!} \int_0^t (t-s)^n \Gamma_x(s) \Gamma^n_{(x)} f ds, \qquad [2.52]$$

$\forall x \in X, \forall f \in \bigcap_x Dom(\Gamma^n(x))$. Taking into account [2.48]–[2.52], we obtain the above-mentioned results.

To be more precise, we have supposed that the following conditions are satisfied everywhere:

A) there exist Hilbert spaces H and H^* compactly imbedded in Banach spaces B and B^*, respectively, $H \subset B, \quad H^* \subset B^*$, where B^* is a dual space to B, that divides points of B;

B) operators $\Gamma(x)$ and$(\Gamma(x))^*$ are dissipative on any Hilbert space H and H^*, respectively;

C) operators $D(x,y)$ and $D^*(x,y)$ are contractive on any Hilbert space H and H^*, respectively;

D) $(x_n;\, n \geq 0)$ is a uniformly ergodic Markov chain with stationary distribution $\rho(A),\, A \in \mathcal{X}$;

E) $m_i(x) := \int_0^\infty t^i G_x(dt)$ are uniformly integrable, $\forall\, i = 1, 2, 3$, where

$$G_x(t) := P\{\omega : \theta_{n+1} \leq t | \, x_n = x\}; \qquad [2.53]$$

F) $\int_X \rho(dx) \|\Gamma(x) f\|^k < +\infty; \quad \int_X \rho(dx) \|\mathrm{P} D_j(x, \cdot) f\|^k < +\infty;$

$$\int_X \rho(dx) \|\Gamma(x) f\|^{k-1} \cdot \|\mathrm{P} D_j(x, \cdot) f\|^{k-1} < +\infty; \qquad [2.54]$$

$\forall k = 1,2,3,4,\ f \in B$, where P is an operator generated by the transition probabilities $P(x, A)$ of Markov chain $(x_n;\ n \geq 0)$:

$$P(x, A) := P\{\omega : x_{n+1} \in A |\, x_n = x\}, \tag{2.55}$$

and $\{D_j(x, y);\ x, y \in X,\ j = 1, 2\}$ is a family of some closed operators.

If $B := C_0(\mathbb{R})$, then $H := W^{l,2}(\mathbb{R})$ is a Sobolev space (Sobolev 1991), and $W^{l,2}(\mathbb{R}) \subset C_0(\mathbb{R})$ and this imbedding is compact. It is the same for the spaces $B := L_2(\mathbb{R})$ and $H := W^{l,2}(\mathbb{R})$.

It follows from the conditions (A) and (B) that operators $\Gamma(x)$ and $(\Gamma(x))^*$ generate a strongly continuous contractive semigroup of operators $\Gamma_x(t)$ and $\Gamma_x^*(t)$, $\forall x \in X$, in H and H^*, respectively. From the conditions (A)–(C), it follows that SMRE $V(t)$ in (1) is a contractive operator in H, $\forall t \in \mathbb{R}_+$, and $\|V(t)f\|_H$ is a semimartingale $\forall f \in H$. Therefore, the conditions (A)–(C) supply the following result:

SMRE $V(t)f$ is a tight process in B. Namely, $\forall \Delta > 0$ there exists a compact set K_Δ

$$\mathrm{P}\{V(t)f \in K_\Delta;\ 0 \leq t \leq T\} \geq 1 - \Delta. \tag{2.56}$$

This result follows from the Kolmogorov–Doob inequality (Jacod and Shiryaev 2010) for semimartingale $\|V(t)f\|_H$ (Korolyuk and Swishchuk 1995b).

Condition [2.56] is the main step in proving limit theorems and rates of convergence for the sequence of SMRE in a series scheme.

2.2.2. *Averaging of HRE*

Let us consider a SMRE in a series scheme:

$$V_\varepsilon(t) = f + \int_0^t \Gamma(x(s/\varepsilon)) V_\varepsilon(s) f ds$$
$$+ \sum_{k=1}^{\nu(t/\varepsilon)} \left[D^\varepsilon(x_{k-1}, x_k) - I\right] V_\varepsilon(\varepsilon\tau_k^-) f, \tag{2.57}$$

where

$$D^\varepsilon(x, y) = I + \varepsilon D_1(x, y) + O(\varepsilon), \tag{2.58}$$

$\{D_1(x,y);\, x,y \in X\}$ is a family of closed linear operators, $\|O(\varepsilon)f\|/\varepsilon \to 0$ as $\varepsilon \to 0$, and ε is a small parameter,

$$f \in B_0 := \bigcap_{x,y\in X} Dom(\Gamma^2(x)) \cap Dom(D_1^2(x,y)). \qquad [2.59]$$

Another form for $V_\varepsilon(t)$ in [2.57] is

$$V_\varepsilon(t) = \Gamma_{x(t/\varepsilon)}(t - \varepsilon\tau_{\nu(t/\varepsilon)}) \prod_{k=1}^{\nu(t/\varepsilon)} D^\varepsilon(x_{k-1}, x_k)\Gamma_{k-1}(\varepsilon\theta_k). \qquad [2.60]$$

Under conditions (A)–(C), the sequence of SMRE $V_\varepsilon(t)f$ is tight (see [2.56]), $\rho - a.s.$

Under conditions $(D,E)\, i = 2$, $(F) k = 2, j = 1$, the sequence of SMRE $V_\varepsilon(t)f$ is weakly compact $\rho - a.s.$ in $D_B[0,+\infty)$ with limit points in $C_B[0,+\infty)$, $f \in B_0$.

Let us consider the following process in $D_B[0,+\infty)$:

$$M^\varepsilon_{\nu(t/\varepsilon)} f^\varepsilon := V^\varepsilon_{\nu(t/\varepsilon)} f^\varepsilon - f^\varepsilon - \sum_{k=0}^{\nu(t/\varepsilon)-1} \mathbf{E}_\rho[(V^\varepsilon_{k+1} f^\varepsilon_{k+1} - V^\varepsilon_k f^\varepsilon_k)|\, \mathcal{F}_k], \qquad [2.61]$$

where $V^\varepsilon_n := V_\varepsilon(\varepsilon\tau_n)$ (see [2.19]),

$$f^\varepsilon := f + \varepsilon f_1(x(t/\varepsilon)), \qquad f^\varepsilon_k := f^\varepsilon(x_k),$$

function $f_1(x)$ is defined from the equation

$$(\mathrm{P} - I)f_1(x) = \Big[(\hat\Gamma + \hat D) - (m(x)\Gamma(x) + \mathrm{P}D_1(x,\cdot))\Big]\, f, \qquad [2.62]$$

$$\hat\Gamma := \int_x \rho(dx)m(x)\Gamma(x), \quad \hat D := \int_x \rho(dx)\mathrm{P}D_1(x,\cdot), \quad m(x) := m_1(x)$$

(see E), $f \in B_0$.

The process $M^\varepsilon_{\nu(t/\varepsilon)} f^\varepsilon$ is an $\mathcal{F}^\varepsilon_t$-martingale with respect to the σ-algebra $\mathcal{F}^\varepsilon_t := \sigma\{x(s/\varepsilon);\, 0 \le s \le t\}$.

The martingale $M^\varepsilon_{\nu(t/\varepsilon)} f^\varepsilon$ in [2.61] has the asymptotic representation:

$$M^\varepsilon_{\nu(t/\varepsilon)} f^\varepsilon = V^\varepsilon_{\nu(t/\varepsilon)} f - f - \varepsilon \sum_{k=0}^{\nu(t/\varepsilon)} (\hat\Gamma + \hat D) V^\varepsilon_k f + O_f(\varepsilon), \qquad [2.63]$$

where $\hat{\Gamma}, \hat{D}, f, f^\varepsilon$ are defined in [2.61]–[2.62] and

$$\|O_f(\varepsilon)\|/\varepsilon \to \text{const} \quad \text{as } \varepsilon \to 0, \quad \forall f \in B_0.$$

We have used [2.19] and [2.20] as $n = 2$, and representations [2.51] and [2.61] in [2.63].

The families $l(M^\varepsilon_{\nu(t/\varepsilon)} f^\varepsilon)$ and

$$l\left(\sum_{k=0}^{\nu(t/\varepsilon)} \mathbf{E}_\rho[(V^\varepsilon_{k+1} f^\varepsilon_{k+1} - V^\varepsilon_k f^\varepsilon_k) | \mathcal{F}_k]\right)$$

are weakly compact for all $l \in B_0^*$, where B_0^* is some dense subset from B^*.

Let $V_0(t)$ be a limit process for $V_\varepsilon(t)$ as $\varepsilon \to 0$.

Since (see [2.60])

$$[V_\varepsilon(t) - V^\varepsilon_{\nu(t/\varepsilon)}] = [\Gamma_{x(t/\varepsilon)}(t - \varepsilon\tau_{\nu(t/\varepsilon)}) - I] \cdot V^\varepsilon_{\nu(t/\varepsilon)} \qquad [2.64]$$

and the right-hand side in [2.64] tends to zero as $\varepsilon \to 0$, then it is clear that the limits for $V_\varepsilon(t)$ and $V^\varepsilon_{\nu(t/\varepsilon)}$ are the same, i.e. $V_0(t) \quad \rho - a.s.$

The sum $\varepsilon \cdot \sum_{k=0}^{\nu(t/\varepsilon)} (\hat{\Gamma} + \hat{D}) V^\varepsilon_k f$ converges strongly as $\varepsilon \to 0$ to the integral

$$\frac{1}{m} \int_0^t (\hat{\Gamma} + \hat{D}) V_0(s) f ds.$$

The quadratic variation of the martingale $l(M^\varepsilon_{\nu(t/\varepsilon)} f^\varepsilon)$ tends to zero, and

$$M^\varepsilon_{\nu(t/\varepsilon)} f^\varepsilon \to 0 \quad \text{as} \quad \varepsilon \to 0, \quad \forall f \in B_0, \quad \forall l \in B_0^*.$$

Passing to the limit in [2.63] as $\varepsilon \to 0$ and taking into account all the previous reasoning, we obtain that the limit process $V_0(t)$ satisfies the equation

$$0 = V_0(t) f - f - \frac{1}{m} \int_0^t (\hat{\Gamma} + \hat{D}) V_0(s) f ds, \qquad [2.65]$$

where

$$m := \int_X \rho(dx) m(x), \quad f \in B_0, \quad t \in [0, T].$$

2.2.3. *Diffusion approximation of HRE*

Let us consider SMRE $V_\varepsilon(t/\varepsilon)$, where $V_\varepsilon(t)$ is defined in [2.57] or [2.60], with the operators

$$D^\varepsilon(x,y) := I + \varepsilon D_1(x,y) + \varepsilon^2 D_2(x,y) + O(\varepsilon^2), \qquad [2.66]$$

$\{Di(x,y);\ x,y \in X,\ i = 1,2\}$ are closed linear operators and $\|O(\varepsilon^2)f\|/\varepsilon^2 \to 0$, as $\varepsilon \to 0$

$$\forall\, f \in B_0 := \bigcap_{x,y\in X} Dom(\Gamma^4(x)) \bigcap Dom(D_2(x,y)),$$

$$Dom(D_2(x,y)) \subseteq Dom(D_1(x,y)); \quad D_1(x,y) \subseteq Dom(D_1(x,y)),$$
$$\forall\, x,y \in X, \quad \Gamma^i(x) \subset Dom(D_2(x,y)), \quad i = 1,2,3. \quad [2.67]$$

Hence,

$$V_\varepsilon(t/\varepsilon) = \Gamma_{x(t/\varepsilon^2)}(t/\varepsilon - \varepsilon\tau_{\nu(t/\varepsilon^2)}) \prod_{k=1}^{\nu(t/\varepsilon^2)} D^\varepsilon(x_{k-1},x_k)\Gamma_{x_{k-1}}(\varepsilon,\theta_k), \ [2.68]$$

where $D^\varepsilon(x,y)$ are defined in [2.66].

Under conditions (A)–(C), the sequence of SMRE $V_\varepsilon(t/\varepsilon)f$ is tight (see [2.56]) $\rho - a.s.$

Under conditions (D, E), $i = 3$, $(F)\ k = 4$, the sequence of SMRE $V_\varepsilon(t/\varepsilon)f$ is weakly compact $\rho - a.s.$ in $D_B[0,+\infty)$ with limit points in $C_B[0,+\infty)$, $f \in B_0$.

Let us assume that the **balance condition** is satisfied, then

$$\int_X \rho(dx)[m(x)\Gamma(x) + \mathrm{P}D_1(x,\cdot)]f = 0, \quad \forall\, f \in B_0. \qquad [2.69]$$

And let us consider the following process in $D_B[0,+\infty)$,

$$M^\varepsilon_{\nu(t/\varepsilon^2)} f^\varepsilon := V^\varepsilon_{\nu(t/\varepsilon^2)} f^\varepsilon - f^\varepsilon - \sum_{k=0}^{\nu(t/\varepsilon^2)-1} \mathbf{E}_\rho[V^\varepsilon_{k+1} f^\varepsilon_{k+1} - V^\varepsilon_k f^\varepsilon_k | \mathcal{F}_k], \ [2.70]$$

where $f^\varepsilon := f + \varepsilon f_1(x(t/\varepsilon^2)) + \varepsilon^2 f_2(x(t/\varepsilon^2))$, and functions f_1 and f_2 are defined from the following equations:

$$(\mathrm{P} - I)f_1(x) = -[m(x)\Gamma(x) + \mathrm{P}D_1(x,\cdot)]f, \qquad [2.71]$$

$$(\mathrm{P}-I)f_2(x)=[\hat{L}-L(x)]f, \qquad \hat{L}:=\int_X \rho(dx)L(x), \qquad [2.72]$$

$$L(x):=(m(x)\Gamma(x)+\mathrm{P}D_1(x,\cdot))(R_0-I)(m(x)\Gamma(x)+\mathrm{P}D_1(x,\cdot))$$
$$+m_2(x)\Gamma^2(x)/2+m(x)\mathrm{P}D_1(x,\cdot)\Gamma(x)+\mathrm{P}D_2(x,\cdot),$$

R_0 is a potential operator of $(x_n;\ n\geq 0)$.

The balance condition [2.69] and condition $\prod(\hat{L}-L(x))=0$ give the solvability of the equations in [2.71].

The process $M^\varepsilon_{\nu(t/\varepsilon^2)}f^\varepsilon$ is an $\mathcal{F}^\varepsilon_t$-martingale with respect to the σ-algebra $\mathcal{F}^\varepsilon_t := \sigma\{x(s/\varepsilon^2);\ 0\leq s\leq t\}$.

This martingale has the asymptotic representation:

$$M^\varepsilon_{\nu(t/\varepsilon^2)}f^\varepsilon = V^\varepsilon_{\nu(t/\varepsilon^2)}f - f - \varepsilon^2\sum_{k=0}^{\nu(t/\varepsilon^2)-1}\hat{L}V^\varepsilon_k f - O_f(\varepsilon t), \qquad [2.73]$$

where $\hat{L}$ is defined in [2.72] and

$$\|O_f(\varepsilon)\|/\varepsilon\to \mathrm{const} \quad \varepsilon\to 0, \quad \forall f\in B_0.$$

We have used [2.19] and [2.20] with $n=3$, and representations [2.70] and [2.71] in [2.72].

The families $l(M^\varepsilon_{\nu(t/\varepsilon^2)}f^\varepsilon)$ and $l(\sum_{k=0}^{\nu(t/\varepsilon^2)}\mathbf{E}_\rho[(V^\varepsilon_{k+1}f^\varepsilon_{k+1} - V^\varepsilon_k f^\varepsilon_k)|\,\mathcal{F}_k])$ are weakly compact for all $l\in B_0^*$, $f\in B_0$.

Let us assume that $V^0(t)$ is the limit process for $V_\varepsilon(t/\varepsilon)$ as $\varepsilon\to 0$.

From [2.60], we obtain that the limits for $V_\varepsilon(t/\varepsilon)$ and $V^\varepsilon_{\nu(t/\varepsilon^2)}$ are the same, namely, $V^0(t)$.

The sum $\varepsilon^2\sum_{k=0}^{\nu(t/\varepsilon^2)}\hat{L}V^\varepsilon_k f$ converges strongly as $\varepsilon\to 0$ to the integral $\frac{1}{m}\int_0^t \hat{L}V^0(s)f ds$.

Now, let us suppose that $M^0(t)f$ is the limit martingale for $M^\varepsilon_{\nu(t/\varepsilon^2)}f^\varepsilon$ as $\varepsilon\to 0$.

Then, from [2.72] to [2.73] and previous reasoning, we have as $\varepsilon\to 0$:

$$M^0(t)f = V^0(t)f - f - \frac{1}{m}\int_0^t \hat{L}V^0(s)f ds. \qquad [2.74]$$

The quadratic variation of the martingale $M^0(t)f$ is of the form

$$< l(M^0(t)f) >= \int_0^t \int_X l^2 \left(\sigma(x)\Gamma(x)V^0(s)f\right) \sigma(dx)ds, \qquad [2.75]$$

where

$$\sigma^2(x) := [m_2(x) - m^2(x)]/m.$$

The solution of the martingale problem for $M^0(t)$ (i.e. to find the representation of $M^0(t)$ with quadratic variation [2.74]) is expressed by the integral over the Wiener orthogonal martingale measure $W(dx, ds)$ with quadratic variation $\rho(dx) \cdot ds$:

$$M^0(t)f = \int_0^t \int_x \sigma(x)\Gamma(x)V^0(s)fW(dx, ds). \qquad [2.76]$$

Hence, the limit process $V^0(t)$ satisfies the following equation (see [2.74] and [2.75]):

$$V^0(t)f = f + \frac{1}{m}\int_0^t \hat{L}V^0(s)fds + \int_0^t \int_X \sigma(x)\Gamma(x)V^0(s)fW(dx, ds). \qquad [2.77]$$

If the operator $\hat{L}$ generates the semigroup $U(t)$, then the process $V^0(t)f$ in Cox (1976) satisfies equation:

$$V^0(t)f = U(t)f + \int_0^t \int_x \sigma(x)U(t-s)\Gamma(x)V^0(s)fW(dx, ds). \qquad [2.78]$$

The **uniqueness** of the limit evolution $V_0(t)f$ in the **averaging** scheme follows from equation [2.78] and the fact that if the operator $\hat{\Gamma} + \hat{D}$ (see [2.62]) generates a semigroup, then $V_0(t)f = \exp\{(\hat{\Gamma} + \hat{D}) \cdot t\}f$ and this representation is unique.

The **uniqueness** of the limit evolution $V^0(t)f$ in **diffusion approximation** scheme follows from the uniqueness of the solution of martingale problem for $V^0(t)f$ (see [2.74]–[2.75]) (Stroock and Varadhan 1969). The latter is proved by **dual SMRE** in the series scheme by constructing the limit equation in diffusion approximation and by using a dual identify (Korolyuk and Swishchuk 1995b).

2.2.4. *Averaging of REs in reducible phase space: merged HRE*

Suppose that the following conditions hold true:

a) **decomposition** of phase space X (**reducible** phase space):

$$X = \bigcup_{u \in U} X_u, \quad X_u \bigcap X_{u'} = \emptyset, \quad u \neq u' : \qquad [2.79]$$

where $(U, \mathcal{U})$ is some measurable phase space (**merged** phase space);

b) a Markov renewal process $(x_n^\varepsilon, \theta_n; n \geq 0)$ on $(X, \mathcal{X})$ has the **semi-Markov kernel**:

$$Q_\varepsilon(x, A, t) := P_\varepsilon(x, A) G_x(t), \qquad [2.80]$$

where $P_\varepsilon(x, A) = P(x, A) - \varepsilon^l P_1(x, A)$, $x \in X$, $A \in \mathcal{X}$, $= \infty, \in$; $P(x, A)$ are the transition probabilities of the **supporting non-perturbed** Markov chain $(x_n;\ n \geq 0)$;

c) the stochastic kernel $P(x, A)$ is adapted to the decomposition (Da Fonseca *et al.* 2009) in the following form:

$$P(x, X_u) = \begin{cases} 1, & x \in X_u \\ 0, & x \notin X_u, \quad u \in U; \end{cases}$$

d) the Markov chain $(x_n; n \geq 0)$ is uniformly ergodic with stationary distributions $\rho_u(B)$:

$$\rho_u(B) = \int_{X_u} P(x, B) \rho_u(dx), \quad \forall u \in U, \quad \forall B \in \mathcal{X}. \qquad [2.81]$$

e) there is a family $\{\rho_u^\varepsilon(A); u \in U, A \in \chi, \varepsilon > 0\}$ of stationary distributions of perturbed Markov chain $(x_n^\varepsilon; n \geq 0)$;

f) $$b(u) := \int_{X_u} \rho_u(dx) P_1(x, X_u) > 0, \quad \forall u \in U,$$

$$b(u, \Delta) := -\int_{X_u} \rho_u(dx) P_1(x, X_\Delta) > 0, \quad \forall u \notin \Delta, \quad \Delta \in U; \qquad [2.82]$$

g) the operators $\Gamma(u) := \int_{X_u} \rho_u(dx) m(x) \Gamma(x)$, and

$$\hat{D}(u) := \int_{X_u} \rho_u(dx) \int_{X_u} P(x, dy) D_1(x, y) \qquad [2.83]$$

are closed $\forall u \in U$ with common domain B_0, and operators $\hat{\Gamma}(u) + \hat{D}(u)$ generate the semigroup of operators $\forall u \in U$.

Decomposition [2.79] in (a) defines the **merging** function

$$u(x) = u, \quad \forall x \in Xu, \quad u \in U. \qquad [2.84]$$

Note that σ-algebras $\mathcal{X}$ and $\mathcal{U}$ are coordinated such that

$$X_\Delta = \bigcup_{u \in \Delta} Xu, \quad \forall u \in U, \quad \Delta \in \mathcal{U}. \qquad [2.85]$$

We set $\prod_u f(u) := \int_{X_u} \rho_u(dx) f(x)$ and $x^\varepsilon(t) := x^\varepsilon_{\nu(t/\varepsilon^2)}$.

SMRE in reducible phase space X is defined by the solution of the equation:

$$\begin{aligned} V_\varepsilon(t) &= I + \int_0^t \Gamma(x^\varepsilon(s/\varepsilon)) V_\varepsilon(s) ds \\ &\quad + \sum_{k=0}^{\nu(t/\varepsilon)} [D^\varepsilon(x^\varepsilon_{k-1}, x^\varepsilon_k) - I] V_\varepsilon(\varepsilon \tau_k^-), \end{aligned} \qquad [2.86]$$

where $D^\varepsilon(x, y)$ are defined in [2.58].

Let us consider the martingale

$$\begin{aligned} M^\varepsilon_{\nu(t/\varepsilon)} f^\varepsilon(x^\varepsilon(t/\varepsilon)) &:= V^\varepsilon_{\nu(t/\varepsilon)} f^\varepsilon(x^\varepsilon(t/\varepsilon)) - f^\varepsilon(x) \\ &\quad - \sum_{k=0}^{\nu(t/\varepsilon)-1} \mathbf{E}_{\rho^\varepsilon_u} [V^\varepsilon_{k+1} f^\varepsilon_{k+1} - V^\varepsilon_k f^\varepsilon_k | \mathcal{F}^\varepsilon_k], \end{aligned} \qquad [2.87]$$

where

$$\mathcal{F}^\varepsilon_n := \sigma\{x^\varepsilon_k, \theta_k; 0 \le k \le n\},$$

$$f^\varepsilon(x) := \hat{f}(u(x)) + \varepsilon f^1(x), \quad \hat{f}(u) := \int_{X_u} \rho_u(dx) f(x), \qquad [2.88]$$

$$\begin{aligned} (\mathrm{P} - I) f_1(x) &= [-(m(x)\Gamma(x) + \mathrm{P}D_1(x, \cdot)) + \hat{\Gamma}(u) \\ &\quad + \hat{D}(u) + (\Pi_u - I) P_1] \hat{f}(u), \\ f^\varepsilon_k := f^\varepsilon(x^\varepsilon_k), &\quad V^\varepsilon_n := V_\varepsilon(\varepsilon \tau_n), \end{aligned} \qquad [2.89]$$

and $V_\varepsilon(t)$ is defined in [2.86], and P_1 is an operator generated by $P_1(x, A)$ (see [2.80]).

The following representation is true (Korolyuk and Swishchuk 1995b):

$$\Pi_u^\varepsilon = \Pi_u - \varepsilon^r \Pi_u P_1 R_0 + \varepsilon^{2r} \Pi_u^\varepsilon (P_1 R_0)^2, \quad r = 1, 2, \qquad [2.90]$$

where $\prod_u^\varepsilon$, $\prod_u$, P_1 are the operators generated by ρ_u^ε, ρ_u and $P_1(x, A)$, respectively, $x \in X$, $A \in \mathcal{X}$, $u \in U$.

It follows from Di Crescenzo (2001) that for any continuous and bounded function $f(x)$

$$\mathbf{E}_{\rho_u^\varepsilon}[f(x)] \to \mathbf{E}_{\rho_u}[f(x)] \quad \text{as} \quad \varepsilon \to 0, \quad \forall u \in U,$$

and the whole calculation in section 2.2.2 can be used in this section after replacing $\mathbf{E}_{\rho_u}$ by $\mathbf{E}_{\rho_u^\varepsilon}$, and it reduces the calculations by $\mathbf{E}_{\rho_u}$ as $\varepsilon \to 0$.

Under conditions (A) $- -$(C), the sequence of SMRE $V_\varepsilon(t)f$ in [2.86], $f \in B_0$, is tight $\rho_u - a.s., \quad \forall u \in U$.

Under conditions (D, E), $i = 2$, (F), $k = 2$, $j = 1$, the sequence of SMRE $V_\varepsilon(t)f$ is weakly compact $\rho_u - a.s.$, $\forall u \in U$, in $D_B[0, +\infty)$ with limit points in $C_B[0, +\infty)$.

Note that $u(x^\varepsilon(t/\varepsilon)) \to \hat{x}(t)$ as $\varepsilon \to 0$, where $\hat{x}(t)$ is a **merged** jump Markov process in $(U, \mathcal{U})$ with **infinitesimal operator** $\Lambda(\hat{P} - I)$,

$$\Lambda \hat{f}(u) := [b(u)/m(u)]\hat{f}(u), \qquad \hat{P}\hat{f}(u) := \int_U [b(u, du')/b(u)]\hat{f}(u), \qquad [2.91]$$

$$m(u) := \int_{X_u} \rho_u(dx) m(x),$$

$b(u)$ and $b(u, \Delta)$ are defined in [2.82]. We also note that

$$\Pi_u P_1 = \Lambda(\hat{P} - I), \qquad [2.92]$$

where $\prod_u$ is defined in [2.90], P_1 in [2.90], Λ and $\hat{P}$ in [2.91].

Using [2.19] and [2.20] as $n = 2$, and [2.88]–[2.89] and [2.90] as $r = 1$, [2.92], we obtain the following representation:

$$M_{\nu(t/\varepsilon)}^\varepsilon f^\varepsilon(x^\varepsilon(t/\varepsilon)) = V_{\nu(t/\varepsilon)}^\varepsilon \hat{f}(u(x^\varepsilon(t/\varepsilon))) - \hat{f}(u(x)) \qquad [2.93]$$

$$-\varepsilon \sum_{k=0}^{\nu(t/\varepsilon)} [m(u)\hat{\Gamma}(u) + m(u)\hat{D}(u) + m(u)\Lambda(\hat{P} - I)] V_k^\varepsilon \hat{f}(u(x_k^\varepsilon)) + O_f(\varepsilon),$$

where $\|O_f(\varepsilon)\|/\varepsilon \to \text{const } \varepsilon \to 0, \quad \forall f \in B_0$. Since the third term in [2.93] tends to the integral

$$\int_0^t [\Lambda(\hat{P} - I) + \hat{\Gamma}(\hat{x}(s)) + \hat{D}(\hat{x}(s))] \times \hat{V}_0(s)\hat{f}(\hat{x}(s))ds$$

and the quadratic variation of the martingale $l(M^{\varepsilon}_{\nu(t/\varepsilon)} f^{\varepsilon}(x^{\varepsilon}(t/\varepsilon)))$ tends to zero as $\varepsilon \to 0$ (hence, $M^{\varepsilon}_{\nu(t/\varepsilon)} f^{\varepsilon}(x^{\varepsilon}(t/\varepsilon)) \to 0, \varepsilon \to 0$), $\forall l \in B_0^*$, then we obtain from [2.93] that the limit evolution $\hat{V}_0(t)$ satisfies equation

$$\hat{V}_0(t)\hat{f}(\hat{x}(t)) = \hat{f}(u) + \int_0^t [\Lambda(\hat{P} - I) + \hat{\Gamma}(\hat{x}(s)) + \hat{D}(\hat{x}(s))]\hat{V}_0(s)\hat{f}(\hat{x}(s))ds. \quad [2.94]$$

The RE $\hat{V}_0(t)$ is called a **merged RE** in the averaging scheme.

2.2.5. *Diffusion approximation of HRE in reducible phase space*

Let us consider SMRE $V_\varepsilon(t/\varepsilon)$ with expansion [2.66], where $V_\varepsilon(t)$ is defined in [2.86], and conditions (A) $- -$(F) (as $i = 3, k = 4, j = 1, 2$) and conditions (a) $- -$(f)$(e = 2)$ are satisfied.

Let us state that the balance condition

$$\int_{X_u} \rho_u(dx)[m(x)\Gamma(x) + \mathrm{P}D_1(x, \cdot)]f = 0, \quad \forall u \in U, \quad [2.95]$$

is also satisfied and operator

$$L(u) := \int_{X_u} \rho_u(dx)L(x)/m(u), \quad [2.96]$$

generates the semigroup of operators, where $L(x)$ is defined in [2.71] and $m(u)$ in [2.91].

Now, consider the martingale

$$\begin{aligned} M^{\varepsilon}_{\nu(t/\varepsilon^2)} f^{\varepsilon}(x^{\varepsilon}(t/\varepsilon^2)) &= V^{\varepsilon}_{\nu(t/\varepsilon^2)} f^{\varepsilon}(x^{\varepsilon}(t/\varepsilon^2)) - f^{\varepsilon}(x) \\ &\quad - \sum_{k=0}^{\nu(t/\varepsilon^2)} \mathbf{E}_{\rho^{\varepsilon}_u}[V^{\varepsilon}_{k+1} f^{\varepsilon}_{k+1} - V^{\varepsilon}_k f^{\varepsilon}_k | \mathcal{F}^{\varepsilon}_k], \end{aligned} \quad [2.97]$$

where

$$\begin{aligned} f^\varepsilon(x) &:= \hat{f}(u(x)) + \varepsilon f^1(x) + \varepsilon^2 f^2(x), \\ (\mathrm{P} - I)f^1(x) &= [m(x)\Gamma(x) + \mathrm{P}D_1(x,\cdot)]\hat{f}(u), \\ (\mathrm{P} - I)f^2(x) &= [m(u)L(u) - L(x) + (\Pi_u - I)P_1]\hat{f}(u), \end{aligned} \qquad [2.98]$$

and $L(u)$ is defined in [2.96].

From the balance condition [2.95] and from the condition

$$\Pi_u[L(u) - L(x) + (\Pi_u - I)P_1] = 0$$

it follows that functions $f^i(x)$, $i = 1, 2$, are defined to be unique.

Assume that $\hat{V}^0(t)$ is the limit of $V_\varepsilon(t/\varepsilon)$ as $\varepsilon \to 0$. Thus, from [2.64] the limits for $V_\varepsilon(t/\varepsilon)$ and $V^\varepsilon_{\nu(t/\varepsilon^2)}$ can be obtained and they are the same, namely, $\hat{V}^0(t)$.

Weakly compactness of $V_\varepsilon(t/\varepsilon)$ is analogue to the one in section 2.2.3 with the use of [2.80] as $l = 2$ and [2.89] as $r = 2$. That is why the whole calculations in section 2.2.3 can be used in this section just replacing $\mathbf{E}_{\rho_u}$ by $\mathbf{E}_{\rho_u^\varepsilon}$, and it reduces to the rates by $\mathbf{E}_{\rho_u}$ as $\varepsilon \to 0$.

Using [2.19] and [2.20] with $n = 3$, as well as representations [2.66] and [2.97]–[2.98], we have the following representation for $M^\varepsilon f^\varepsilon$:

$$\begin{aligned} &M^\varepsilon_{\nu(t/\varepsilon^2)} f^\varepsilon = V^\varepsilon_{\nu(t/\varepsilon^2)} \hat{f}(u(x^\varepsilon(t/\varepsilon^2))) - \hat{f}(u)(x) \\ &-\varepsilon^2 \sum_{k=0}^{\nu(t/\varepsilon^2)} [m(u)L(u(x^\varepsilon_k)) + \Pi_u P_1] V^\varepsilon_k \hat{f}(u(x^\varepsilon_k)) + O_f(\varepsilon), \end{aligned} \qquad [2.99]$$

where $L(u)$ is defined in [2.96], and $\|O_f(\varepsilon)\|/\varepsilon \to \text{const}$ as $\varepsilon \to 0$. The sum in [2.99] converges strongly as $\varepsilon \to 0$ to the integral

$$\int_0^t [\Lambda(\hat{P} - I) + L(\hat{x}(s))]\hat{V}^0(s)\hat{f}(\hat{x}(s))ds, \qquad [2.100]$$

because of relation [2.91], where $\hat{x}(t)$ is a jump Markov process in (U, U) with infinitesimal operator $\Lambda(\hat{P} - I)$, $\hat{x}(0) = u \in U$.

Let $\hat{M}^0(t)f$ be a limit martingale for

$$M^\varepsilon_{\nu(t/\varepsilon^2)} f^\varepsilon(x^\varepsilon(t/\varepsilon^2)) \text{ as } \varepsilon \to 0.$$

From [2.94]–[2.99], we have the equation, as $\varepsilon \to 0$,

$$\hat{M}^0(t)\hat{f}(\hat{x}(t)) = \hat{V}^0(t)\hat{f}(\hat{x}(t)) - \hat{f}(u) - \int_0^t [\Lambda(\hat{P} - I) + L(\hat{x}(s))]\hat{V}^0(s)\hat{f}(\hat{x}(s))ds. \quad [2.101]$$

The quadratic variation of the martingale $\hat{M}^0(t)$ has the form

$$< l(\hat{M}^0(t)\hat{f}(u)) >= \int_0^t \int_{X_u} l^2(\sigma(x,u)\Gamma(x)\hat{V}^0(s)\hat{f}(u))\rho_u(dx)ds, \quad [2.102]$$

where

$$\sigma^2(x,u) := [m_2(x) - m^2(x)]/m(u).$$

The solution of the martingale problem for $\hat{M}^0(t)$ is expressed by the integral

$$\hat{M}^0(t)\hat{f}(\hat{x}(t)) = \int_0^t \hat{W}(ds, \hat{x}(s))\hat{V}^0(s)\hat{f}(\hat{x}(s)), \quad [2.103]$$

where

$$\hat{W}(t,u)f := \int_{X_u} W_{\rho_u}(t,dx)\sigma(x,u)\Gamma(x)f.$$

Finally, from [2.100] to [2.102], it follows that the limit process $\hat{V}^0(t)$ satisfies the following equation:

$$\hat{V}^0(t)\hat{f}(\hat{x}(t)) = \hat{f}(u) + \int_0^t [\Lambda(\hat{P} - I) + L(\hat{x}(s))]\hat{V}^0(s)\hat{f}(\hat{x}(s))ds + \int_0^t \hat{W}(ds, \hat{x}(s))\hat{V}^0(s)\hat{f}(\hat{x}(s)). \quad [2.104]$$

The RE $\hat{V}^0(t)$ in [2.104] is called a **merged RE** in the diffusion approximation scheme. If the operator $\hat{U}^0(t)$ is a solution of the Cauchy problem

$$\begin{cases} \dfrac{d\hat{U}^0(t)}{dt} = \hat{U}^0(t)L(\hat{x}(t)) \\ \hat{U}^0(0) = I, \end{cases}$$

then the operator process $\hat{V}^0\hat{f}(\hat{x}(t))$ satisfies the equation

$$\hat{V}^0(t)\hat{f}(\hat{x}(t)) = \hat{U}^0(t)\hat{f}(u) + \int_0^t \hat{U}^0(t-s)\Lambda(\hat{P}-I)\hat{V}^0(s)\hat{f}(\hat{x}(s))ds$$
$$+ \int_0^t \hat{U}^0(t-s)\hat{W}(ds, \hat{x}(s))\hat{V}^0(s)\hat{f}(\hat{x}(s)). \quad [2.105]$$

The uniqueness of the limit of the RE $\hat{V}^0(t)$ is established by the dual of SMRE.

2.2.6. *Normal deviations of HRE*

The averaged evolution obtained by averaging and merging schemes can be considered as the first approximation to the initial evolution. The diffusion approximation of the SMRE determines the second approximation to the initial evolution, since the first approximation under balance condition, the averaged evolution, appears to be trivial.

Now, we consider the **double approximation** to SMRE, i.e. both the averaged and the diffusion approximation, provided that the balance condition fails. We introduce the **deviation process** as the normalized difference between the initial and averaged evolutions. In the limit we will obtain the **normal deviations** of the initial SMRE from the averaged one.

We consider the SMRE $V_\varepsilon(t)$ in [2.57] and the averaged evolution $V_0(t)$ in [2.65]. Let us also consider the deviation of the initial evolution $V_\varepsilon(t)f$ from the averaged one $V_0(t)f$, i.e.

$$W_\varepsilon(t)f := \varepsilon^{-1/2} \cdot [V_\varepsilon(t) - V_0(t)]f, \quad \forall f \in B_0. \quad [2.106]$$

Taking into account equations [2.57] and [2.106], we obtain the relation for $W_\varepsilon(t)$

$$W_\varepsilon(t)f = \varepsilon^{-1/2}\int_0^t \left(\Gamma(x(s/\varepsilon)) - \hat{\Gamma}\right) V_\varepsilon(s)fds + \int_0^t \hat{\Gamma}W_\varepsilon(s)fds$$
$$+\varepsilon^{-1/2}\left[V_\varepsilon^d(t) - \int_0^t \hat{D}\, V_0(s)ds\right] f, \quad \forall f \in B_0, \quad [2.107]$$

where

$$V_\varepsilon^d(t)f := \sum_{k=1}^{\nu(t/\varepsilon)} [D^\varepsilon(x_{k-1}, x_k) - I]V_\varepsilon(\varepsilon\tau_k^-)f,$$

and $\hat{\Gamma}$, $\hat{D}$ are defined in [2.62].

If the process $W_\varepsilon(t)f$ has the weak limit $W_0(t)f$ as $\varepsilon \to 0$, then we obtain

$$\int_0^t \hat{\Gamma} W_\varepsilon(s) f ds \to \int_0^t \hat{\Gamma} W_0(s) f ds, \text{ as } \varepsilon \to 0. \quad [2.108]$$

Since the operator $\Gamma(x) - \hat{\Gamma}$ satisfies the balance condition

$$\left(\Pi(\Gamma(x) - \hat{\Gamma})f = 0\right),$$

then the diffusion approximation of the first term in the right-hand side of [2.107] gives

$$\varepsilon^{-1/2} \int_0^t e\left((\Gamma(x(s/\varepsilon)) - \hat{\Gamma})f\right) ds \to l(\sigma_1 f) w(t), \text{ as } \varepsilon \to 0 \quad [2.109]$$

where

$$l^2(\sigma_1 f) = \int_X \rho(dx)\Big[m(x) l((\Gamma(x) - \hat{\Gamma})f)(R_0 - I) m(x) l((\Gamma(x) - \hat{\Gamma})f) + \frac{1}{2} m_2(x) \frac{1}{2}((\Gamma(x) - \hat{\Gamma})f)\Big]/m,$$

$\forall l \in B_0$, $w(t)$ is a standard Wiener process.

Since $\prod(\mathrm{P}D_1(x,\cdot) - \hat{D})f = 0$, then the diffusion approximation of the third term in the right-hand side of Elliott and Swishchuk (2007) gives the following limit:

$$\varepsilon^{-1/2}\, l\left(V_\varepsilon^d(t)f - \int_0^t \hat{D} V_0(s) f ds\right) \to l(\sigma_2 f)\, w(t), \text{ as } \varepsilon \to 0, \quad [2.110]$$

where

$$l^2(\sigma_2 f) := \int_X \rho(dx) l\left((\mathrm{P}D_1(x,\cdot) - \hat{D})f\right)(R_0 - I)\, l\left((\mathrm{P}D_1(x,\cdot) - \hat{D})f\right).$$

The passage to the limit as $\varepsilon \to 0$ in the representation [2.107] by using [2.108]–[2.110] allows us to arrive at the equation for $W_0(t)f$

$$W_0(t)f = \int_0^t \hat{\Gamma} W_0(s) f ds + \sigma f w(t), \quad [2.111]$$

where the variance operator σ is determined from the relation:

$$l^2(\sigma f) := l^2(\sigma_1 f) + l^2(\sigma_2 f), \quad \forall l \in B_0, \quad \forall l \in B_0^*, \quad [2.112]$$

and operators σ_1 and σ_2 are defined in [2.109] and [2.110], respectively.

Thus, the double approximation of the SMRE has the form

$$V_\varepsilon(t)f \approx V_0(t)f + \sqrt{\varepsilon}\, W_0(t)f$$

for small ε, which perfectly fits the standard form of the CLT with non-zero limiting mean value.

2.2.7. *Rates of convergence in the limit theorems for HRE*

The rates of convergence in the averaging and diffusion approximation scheme for the sequence of SMRE are studied in this section.

Averaging scheme: The problem is to estimate the value

$$||\mathbf{E}_\rho[V_\varepsilon(t)f^\varepsilon(x(t/\varepsilon)) - V_0(t)f]||, \quad \forall f \in B_0, \qquad [2.113]$$

where $V_0(t), V_\varepsilon(t), f^\varepsilon, f$ and B_0 are defined in [2.65], [2.57], [2.61] and [2.59], respectively.

We use the following representation:

$$||\mathbf{E}_\rho[V_\varepsilon(t)f^\varepsilon(x(t/\varepsilon)) - V_0(t)f]|| \leq ||\mathbf{E}_\rho[V_\varepsilon(t)f - V_\varepsilon(\tau_{\nu(t/\varepsilon)})f]||$$
$$+||\mathbf{E}_\rho[V_\varepsilon(\tau_{\nu(t/\varepsilon)})f - V_0(t)f]|| + \varepsilon||\mathbf{E}_\rho V_\varepsilon(t)f_1(x(t/\varepsilon))|| \qquad [2.114]$$

that follows from [2.108] and [2.65], [2.61], and [2.59].

For the first term in the right-hand side of [2.114], we obtain (see [2.64] and [2.52] with $n = 2$):

$$||\mathbf{E}_\rho[V_\varepsilon(t)f - V_\varepsilon(\tau_{\nu(t/\varepsilon)}))f]|| \leq \varepsilon\, C_1(T,f), \quad \forall t \in [0,T], \qquad [2.115]$$

where

$$C_1(T,f) := \int_X \rho(dx)[C_0(T,x,f) + C_0^2(T,x,f)],$$
$$C_0(T,x,f) := T\, m_2(x)||\Gamma(x)f||/2m, \quad \forall f \in B_0.$$

For the second term in the right-hand side of [2.114], we have from [2.63] and [2.113] (since $\mathbf{E}_\rho[M^\varepsilon_{\nu(t/\varepsilon)}f^\varepsilon(x(t/\varepsilon)) = 0]$):

$$||\mathbf{E}_\rho[V_\varepsilon(\tau_{\nu(t/\varepsilon)})f - V_0(t)f]|| \leq \varepsilon\, ||\mathbf{E}_\rho[V^\varepsilon_{\nu(t/\varepsilon)} - I]f_1(x(t/\varepsilon))||$$

$$+\varepsilon \left\| \mathbf{E}_\rho \left[\sum_{k=0}^{\nu(t/\varepsilon)-1} (\hat{\Gamma} + \hat{D}) V_k^\varepsilon f - \frac{1}{\varepsilon m} \int_0^1 (\hat{\Gamma} + \hat{D}) V_0(s) f ds \right] \right\|$$

$$+\varepsilon\, C_2(T, f), \qquad [2.116]$$

where the constant $C_2(T, f)$ is expressed by the algebraic sum of

$$\int_X m_i(x) \|\Gamma^i(x) f\| \rho(dx)$$

and

$$\int_X m_i(x) \|\mathrm{P} D_1(x, \cdot) \cdot \Gamma^i(x) f\| \rho(dx), \quad i = 1, 2, \quad f \in B_0,$$

and $\|R_0\|$, which is a potential of Markov chain $(x_n;\, n \geq 0)$.

For the third term in right-hand side of [2.114], we obtain

$$\mathbf{E}_\rho \|f_1(x)\| \leq 2\, C_3(f), \qquad [2.117]$$

where

$$C_3(f) := \|R_0\| \int_X p(dx) \left[m(x) \|\Gamma(x) f\| + \|\mathrm{P} D_1(x, \cdot) f\| \right].$$

Finally, from [2.114] to [2.117], we obtain the estimate of the value in [2.113], namely, **rate of convergence in averaging scheme for SMRE**:

$$\|\mathbf{E}_\rho [V_\varepsilon(t) f^\varepsilon(x(t/\varepsilon)) - V_0(t) f]\| \leq \varepsilon\, C(T, f), \qquad [2.118]$$

where the constant $C(T, f)$ can be expressed as a function of constants $C_i(T, f)$, $i = 1, 2, 3$.

Diffusion approximation: In this case, the problem is to estimate the value

$$\|\mathbf{E}_\rho [V_\varepsilon(t/\varepsilon) f^\varepsilon(x(t/\varepsilon^2)) - V^0(t) f]\|, \quad \forall f \in B_0, \qquad [2.119]$$

where $V_\varepsilon(t/\varepsilon)$, f^ε, $V^0(t)$, f, and B_0 are defined in [2.68], [2.70], [2.77], and [2.67], respectively.

Here, we use the following representation:

$$\|\mathbf{E}_\rho [V_\varepsilon(t/\varepsilon) f^\varepsilon(x(t/\varepsilon^2)) - V^0(t) f]\| \leq \|\mathbf{E}_\rho [V_\varepsilon(t/\varepsilon) f - V_\varepsilon(\tau_{\nu(t/\varepsilon^2)}) f]\|$$

$$+\|\mathbf{E}_\rho[V_\varepsilon(\tau_{\nu(t/\varepsilon^2)})f - V^0(t)f]\| + \varepsilon\,\|\mathbf{E}_\rho[V_\varepsilon(t/\varepsilon)f_1(x(t/\varepsilon^2))]\|$$
$$+\varepsilon^2\,\|\mathbf{E}_\rho[V_\varepsilon(t/\varepsilon)f_2(x(t/\varepsilon^2))]\|, \qquad [2.120]$$

that follows from [2.118], [2.70] and [2.64], respectively.

First of all, we have for the fourth term in the right-hand side of [2.119]:

$$\varepsilon^2\,\|\mathbf{E}_\rho[V_\varepsilon(t/\varepsilon)f_2(x(t/\varepsilon^2))]\| \le \varepsilon^2\,2\|R_0\|\int_X \rho(dx)\|L(x)f\|$$
$$:= \varepsilon^2\,d_1(f), \qquad [2.121]$$

where $L(x)$ is defined in [2.71].

For the third term in the right-hand side of [2.119], we obtain:

$$\varepsilon\,\|\mathbf{E}_\rho[V_\varepsilon(t/\varepsilon)f_1(x(t/\varepsilon^2))]\| \le \varepsilon\,d_2(f), \qquad [2.122]$$

where

$$d_2(f) := 2\|R_0\|\cdot\int_X \rho(dx)[m(x)\|\Gamma(x)f\| + \|PD_1(x,\cdot)f\|], \quad f\in B_0.$$

For the first term in the right-hand side of [2.119], we have from [2.118]:

$$\|\mathbf{E}_\rho[V_\varepsilon(t/\varepsilon)f - V_\varepsilon(\tau_{\nu(t/\varepsilon^2)})f]\| \le \varepsilon\,C_1(T,f), \qquad [2.123]$$

where $C_1(T,f)$ is defined in [2.118].

For the second term in the right-hand side of [2.124], we use the asymptotic representation [2.72] for the martingale $M^\varepsilon_{\nu(t/\varepsilon^2)}f^\varepsilon$ and the conditions

$$\mathbf{E}_\rho[M^\varepsilon f^\varepsilon] = 0, \quad \mathbf{E}_\rho[M^0(t)f] = 0, \quad \forall f\in B_0 \qquad [2.124]$$

$$\|\mathbf{E}_\rho[V_\varepsilon(\tau_{\nu(t/\varepsilon^2)}))f - V^0(t)f]\| \le \varepsilon\,\|\mathbf{E}_\rho[V^\varepsilon(\tau_{\nu(t/\varepsilon^2)}))f_1 - f_1(x)]\|$$
$$+\varepsilon^2\,\|\mathbf{E}_\rho[V_\varepsilon(\tau_{\nu(t/\varepsilon^2)}))f_2 - f_2(x)]\|$$
$$+\varepsilon^2\left\|\mathbf{E}_\rho\left[\sum_{k=0}^{\nu(t/\varepsilon^2)-1}\hat{L}V_k^\varepsilon f - \frac{1}{\varepsilon^2 m}\int_0^t \hat{L}V^0(s)fds\right]\right\| + \varepsilon\,d_3(f),$$

$$[2.125]$$

where the constant $d_3(f)$ is expressed by the algebraic sum of

$$\int_X m_i(x)\|\Gamma^j(x)PD_e(x,\cdot)f\|\rho(dx), \quad i=1,2,3, \quad j=0,1,2,3, \quad e=1,2.$$

We should note that

$$\left\| \mathbf{E}_\rho \left[\sum_{k=0}^{\nu(t/\varepsilon^2)-1} \hat{L}V_k^\varepsilon f - \varepsilon^{-2}m^{-1} \int_0^t \hat{L}V^0(s)fds \right] \right\| \leq d_4(T,f). \qquad [2.126]$$

Finally, from [2.120] to [2.126], we obtain the estimate of the value in [2.119], namely, **rate of convergence in diffusion approximation scheme for SMRE** as

$$\|\mathbf{E}_\rho[V_\varepsilon(t/\varepsilon)f^\varepsilon(x(t/\varepsilon^2)) - V^0(t)f]\| \leq \varepsilon\, d(T,f), \qquad [2.127]$$

where the constant $d(T,f)$ is expressed by $d_i, i=1,2,3,4$, and $C_1(T,f)$, $f \in B_0$.

PART 2

Applications to Reliability, Random Motions, and Telegraph Processes

3

Asymptotic Analysis for Distributions of Markov, Semi-Markov and Random Evolutions

This chapter is devoted to finding the probabilities of large deviations for trajectories of a semi-Markov process and random evolutions on the line. In particular, we estimate the asymptotic distribution of the reaching time of an infinitely increasing level by a semi-Markov process on the line, as long as a discrete analogue of the action functional for the embedded Markov chain is obtained. We also find the asymptotic estimations of that reaching time if the limit, which defies the discrete analogue of the action functional, does not exist.

In the modern theory of random processes, considerable interest is devoted to problems related to the asymptotic behavior of the instants of achieving "hard to reach" areas of the phase space. Such problems can be divided into the following three mutually complementary groups:

1) *Large deviation theorems*: This type of problems is studied by many research works, including large deviations of empirical measures of Markov processes, reported in fundamental works by Donsker and Varadhan (1975), and large deviations of diffusion processes, studied by Ventzel and Freydlin (1970). A series of works by Borovkov and Mogulsky (1980) and Foss (2007) are devoted to large deviations for sums of random vectors and fields. Large deviations for walks on a multidimensional lattice are studied by Molchanov and Yarovaya (2013), and large deviations for random particle systems are studied by Dawson and Gärtner (1994). This group also includes the recent results of Novak, who investigated the accuracy of the approximation of processes exceeding a threshold level, large deviations for Poisson processes and a Poisson approximation to study extremes (Novak 2007).

2) If a domain D of the phase space is fixed, and some local characteristics of a process depend on a small parameter $\varepsilon > 0$ such that the probability of reaching time at D tends to zero as $\varepsilon \to 0$. Such problems have also been the main focus of many research works. In particular, they were studied by Korolyuk and Turbin (1993), Gnedenko and Ushakov (1995), Skorokhod (1989), Kovalenko (1980), Soloviev (1993), Anisimov (1977), and others.

3) Semi-Markov process on a domain D_ε that depends on a parameter $\varepsilon > 0$ such that the probability to reach the domain by the process tends to zero as $\varepsilon \to 0$. First, we refer to the works of Soloviev (1993) and Vinogradov (1994). Secondly, these problems have also been addressed by Korolyuk and Silvestrov (1983), Korolyuk and Turbin (1993), Silvestrov (2007a, 2007b) and others.

In this chapter, we include some problems of the last group. As in most of the cases in this book, we consider problems for which it is not possible to specify the exact form of the distribution of hard-to-reach states. Therefore, we study the asymptotic distribution. There are multiple applications for these problems. For instance, when we study the probability of losing a customer in a queueing system, in models for storage management, in the theory of runs and so on.

In section 3.1, we deal with the problem of finding an asymptotic distribution of the functional type for reaching a level that is infinitely increasing by a semi-Markov process on the set of natural numbers. For this analysis, the existence of a discrete analogue of the action functional for a Markov chain, embedded in the semi-Markov process, arises as an important condition.

In section 3.2, we find asymptotic estimates for the distribution of the time that a semi-Markov process spends in a set of states. These estimates infinitely expand when the condition for the existence of a discrete analogue of the action functional is not fulfilled.

Section 3.3 is devoted to the asymptotic expansion for the distribution of first exit time from an expanding set of states by the semi-Markov process that is embedded in a diffusion process.

In section 3.4, an asymptotic expansion is obtained for a semigroup of operators of the corresponding three-component Markov process, after the diffusion approximation of the process is stated.

Section 3.5 deals with the asymptotic expansion under the diffusion approximation for the distribution of the position of a particle, which performs a random walk in n-dimensional space with Markov switching, provided Kac's condition is achieved.

3.1. Asymptotic distribution of time to reach a level that is infinitely increasing by a family of semi-Markov processes on the set $\mathbb{N}$

We will obtain the asymptotic expansion for the distribution of a functional type of the time for reaching a level, which is infinitely increasing, by a semi-Markov process with the discrete phase space.

Let $\xi(t)$, $t \geq 0$, be a semi-Markov process on the phase space $E = \{1, 2, 3, \ldots\}$, and its embedded Markov chain ξ_m, $m \in \mathbb{N}$, satisfies the following condition C2. *The Markov chain ξ_m is irreducible and has the stationary distribution $\boldsymbol{\rho} = \{\rho_i,\, i \in E\}$, moreover, $\rho_k > 0$ for any arbitrarily large $k \in E$.*

Denote by e_n a single lumped state (merging a set of states), and by $E^n = \{n+1,\, n+2, \ldots\}$ (Korolyuk *et al.* 1979a,b), i.e. the transition probabilities of the lumped Markov chain $\overline{\xi}_m$, $m \in \mathbb{N}$, on the phase space $E_n \cup \{e_n\}$, where $E_n = \{1, 2, \ldots, n\}$, are of the form of definition 1.12.

We will use the notion of the lumped Markov chain introduced in Chapter 1 of Volume 1: Let us fix a state $k \in E$ and denote by π_{ij}, $i, j \in \{k, e_n\}$, then the transition probabilities of the lumped Markov chain $\widetilde{\xi}_m^{(n)}$, $m \in \mathbb{N}$ (Korolyuk *et al.* 1979b), obtained from the Markov chain ξ_m, are defined on the phase space $E_k^{(n)} = \{k, e_n\}$.

Let us state the following condition:

C2. *There exists*

$$\lim_{n\to\infty} \frac{\rho_k \pi_{ke_n}}{\varepsilon_n} = A > 0,$$

where $\varepsilon_n = \sum_{i>n} \rho_i$.

REMARK 3.1.– *In Korolyuk* et al. *(1979b), it was proved that $0 \leq A \leq 1$ and it was shown that, in some manners, A is a discrete analogue of the action functional, which was introduced by Ventzel and Freydlin (1970).*

On the phase space $E_0 = \{0, 1, 2, \ldots\}$ consider an ergodic irreducible Markov chain $\{\xi_m,\, m \in \mathbb{N}\}$ with the following transition probabilities matrix (this matrix is formed with four submatrices)

$$\begin{pmatrix} [P_0] & \begin{bmatrix} 0 & 0 & \cdots \\ \vdots & \vdots & \ddots \\ p_{k_0-1} & 0 & \cdots \end{bmatrix} \\ \begin{bmatrix} 0 & \cdots & 0 & q_{k_0} \\ 0 & \cdots & 0 & 0 \\ \vdots & \vdots & \vdots & \vdots \end{bmatrix} & \begin{bmatrix} 0 & p_{k_0} & \cdots \\ q_{k_0+1} & 0 & \cdots \\ 0 & q_{k_0+2} & \cdots \\ \vdots & \vdots & \ddots \end{bmatrix} \end{pmatrix}$$

where $[P_0]$ is a $k_0 \times k_0$ submatrix.

Let $\boldsymbol{\rho} = \{\rho_i,\ i \in E_0\}$ be the stationary distribution of $\{\xi_m\}$, and e_n be the new state obtained after merging the set of states $E^n = \{n+1, n+2, \cdots\}$, $\varepsilon_n = \sum_{i>n} \rho_i$.

For a fixed $n_0 < n$, define $E_{n_0} = \{0, 1, \cdots, n_0\}$ and denote by π_{ij}, $i, j \in E_{n_0} \cup e_n$, the transition probabilities of the lumped Markov chain.

LEMMA 3.1.– (Pogorui 1990). Suppose that there exists

$$\lim_{n\to\infty} p_n = p < \frac{1}{2}.$$

Then,

$$A = \frac{(q-p)^2}{q}, \text{ where } \quad q = 1 - p.$$

PROOF.– Since $\boldsymbol{\rho}$ is the stationary distribution of the Markov chain, we have for any $i \geq k_0$

$$\rho_i = \rho_{i-1} p_{i-1} + \rho_{i+1} q_{i+1}.$$

This implies that for $n \geq k_0$

$$\sum_{i\geq n} \rho_i = \sum_{i\geq n} \rho_{i-1} p_{i-1} + \sum_{i\geq n} \rho_{i+1} q_{i+1} = \rho_{n-1} p_{n-1} - \rho_n q_n + \sum_{i\geq n} \rho_i \left(p_i + q_i\right).$$

Therefore, for $n \geq k_0$

$$\frac{\rho_{n-1}}{\rho_n} = \frac{q_n}{p_{n-1}}. \qquad [3.1]$$

It is well-known (Korolyuk *et al.* 1979b) that $A = \lim_{n\to\infty} \sum_{i\in E_{n_0}} \pi_{e_n i}$ and A does not depend on n_0. In addition, if for any $n \geq 1$, $p_n = p$, $q_n = q$, where $q > p$ we have

$$A = \frac{(q-p)^2}{q}.$$

In our case for $n_0 \geq k_0$, we have

$$\sum_{i\in E_{n_0}} \pi_{e_n i} = \pi_{e_n n_0} = \frac{\rho_{n+1} q_{n+1}}{\varepsilon_n} f_{n n_0}^{e_n},$$

where $f_{n n_0}^{e_n}$ is the probability that, after starting from state n, the chain $\{\xi_m\}$ reaches state n_0 before e_n.

Taking into account [3.1], for $n \geq k_0$ we have

$$\frac{\rho_{n+1} q_{n+1}}{\varepsilon_n} = \frac{q_{n+1}}{1 + \frac{p_{n+2}}{q_{n+3}} + \frac{p_{n+2}p_{n+3}}{q_{n+3}q_{n+4}} + \dots}.$$

We need to show that

$$\lim_{n\to\infty} \frac{\rho_{n+1} q_{n+1}}{\varepsilon_n} = q - p.$$

Denote by

$$a_k^n = \prod_{i=1}^{k} \frac{p_{n+i+1}}{q_{n+i+2}}, \quad a_0^n = 1.$$

Hence, $\frac{\varepsilon_n}{\rho_{n+1}} = \sum_{k=0}^{\infty} a_k^n$. Since $p_n \to p < \frac{1}{2}$ as $n \to \infty$ for any fixed $\varepsilon > 0$, there exist N_1 and N_2 such that $\sum_{k=N_1}^{\infty} a_k^n < \frac{\varepsilon}{2}$ for all $n > N_2$. Then, there exists N_3 such that $\left|\sum_{k=0}^{N_1} a_k^n - \sum_{k=0}^{N_1} \left(\frac{p}{q}\right)^k\right| < \frac{\varepsilon}{2}$ for all $n > N_3$.

By defining $N_4 = \max(N_2, N_3)$, we obtain for all $n > N_4$

$$\left|\sum_{k=0}^{\infty} a_k^n - \sum_{k=0}^{\infty} \left(\frac{p}{q}\right)^k\right| < \varepsilon.$$

Therefore,

$$\lim_{n\to\infty} \sum_{k=0}^{\infty} a_k^n = \frac{q}{q-p}.$$

Since $q_n \to q$ as $n \to \infty$, we have $\lim_{n\to\infty} \frac{\rho_{n+1}}{\varepsilon_n} q_{n+1} = q - p$.

Now, it remains to show that $\lim_{n\to\infty} f^{e_n}_{nn_0} = \frac{q-p}{q}$. It is easily verified that $\lim_{n\to\infty} \overline{f}^{e_n}_{nn_0} = \frac{q-p}{q}$, where $\overline{f}^{e_n}_{nn_0}$ is $f^{e_n}_{nn_0}$ in the case when $p_n = p$, $q_n = q$ for all $n \geq k_0$.

Suppose $\varepsilon > 0$ is small such that $q - \varepsilon > p + \varepsilon$. It is easily seen that $\lim_{n\to\infty} f^{e_n}_{nn_0}$ does not depend on n_0 (otherwise A depends on n_0).

Consider N_0 such that $|p_n - p| < \varepsilon$ and $|q_n - q| < \varepsilon$ for any $n > N_0$.

Let $\check{f}^{e_n}_{nn_0}$ be $f^{e_n}_{nn_0}$ in the case where $p_n = p - \varepsilon$, $q_n = q + \varepsilon$ for all $n \geq k_0$ and $\hat{f}^{e_n}_{nn_0}$ be $f^{e_n}_{nn_0}$ in the case where $p_n = p + \varepsilon$, $q_n = q - \varepsilon$ for all $n \geq k_0$. It is then easily verified that

$$\frac{q-p-2\varepsilon}{q-\varepsilon} = \lim_{n\to\infty} \hat{f}^{e_n}_{nn_0} \leq \lim_{n\to\infty} f^{e_n}_{nn_0} \leq \lim_{n\to\infty} \check{f}^{e_n}_{nn_0} = \frac{q-p+2\varepsilon}{q+\varepsilon}.$$

Since $\varepsilon > 0$ can be arbitrary small, we have $\lim_{n\to\infty} f^{e_n}_{nn_0} = \frac{q-p}{q}$.

Now, let us consider the following example of a Markov chain, for which the constant A is calculated directly.

EXAMPLE 3.1.– Let ξ_m, $m \in \mathbb{N}$, be a Markov chain on the phase space $E_0 = \{0, 1, 2, \cdots\}$ with the following transition probabilities:

$$P = \begin{pmatrix} 0 & 1 & 0 & 0 & \cdots \\ q_1 & 0 & p_1 & 0 & \cdots \\ 0 & q_2 & 0 & p_2 & \cdots \\ 0 & 0 & q_3 & 0 & \cdots \\ \vdots & \vdots & \vdots & \vdots & \ddots \end{pmatrix}$$

where $p_i + q_i = 1$ for all $i \geq 1$.

Now suppose that there exists

$$\lim_{n\to\infty} p_n = p < \frac{1}{2}.$$

It was proved in the previous lemma that:

$$A = \frac{(q-p)^2}{q}, \quad where \quad q = 1 - p.$$

Denote by $\tau^{(k)}_n$, $n > k$, the time to reaching the set E^n by $\xi(t)$ assuming $\xi(0) = k$, i.e.

$$\tau^{(k)}_n = \inf\{t : \xi(t) \in E^n, \quad \xi(0) = k\}.$$

We are interested in the asymptotic expansion of the distribution of the random variable $\varepsilon_n \tau_n^{(k)}$ as $n \to \infty$.

Denote by θ_i a sojourn time of $\xi(t)$ at state $i \in E$, and define $a_i = \mathbf{E}[\theta_i]$.

Let the following condition be satisfied:

C3. $\sup_{i \in E} a_i = c < \infty$.

Let $\widetilde{\xi}^{(n)}(t)$, $t \geq 0$, be the lumped semi-Markov process, obtained from $\xi(t)$ and defined on the phase space $E_k^{(n)}$ with the embedded Markov chain $\widetilde{\xi}_m^{(n)}$. Denote by $G_n(t)$ the sojourn time distribution function of $\widetilde{\xi}^{(n)}(t)$ at state k, and we assume that there exists the pdf $g_n(t) = \frac{d}{dt} G_n(t)$. We have removed the explicit state dependence in this cumulative distribution function (cdf) and pdf to ease the notation.

Define the first moment as

$$m_1^{(n)} = \int_0^\infty t\, g_n(t)\, dt.$$

It was shown in Korolyuk *et al.* (1979b) that condition C3 provides the existence of the first moment and

$$\lim_{n\to\infty} m_1^{(n)} = \sum_{k \geq 1} \rho_k a_k.$$

We also have

$$B_n = \frac{\rho_k \pi_k e_n}{m_1^{(n)} \varepsilon_n}, \quad V_n = m_1^{(n)} B_n, \quad R_0^{(n)}(t) = \sum_{i=1}^{\infty} g_n^{(i)*}(t) - \left(m_1^{(n)}\right)^{-1},$$

where

$$g_n^{(i)*}(t) = \int_0^t g_n^{(i-1)*}(t-u)\, g_n(u)\, du,$$

and $g_n^{(1)}(t) = g_n(t)$.

Consider

$$F_{\varepsilon_n}(t) = P\left\{\varepsilon_n \tau_n^{(k)} > t\right\}, \qquad F_0^{(n)}(t) = e^{-B_n t}.$$

LEMMA 3.2.– If the conditions C1–C3 are fulfilled, then

$$F_{\varepsilon_n}(t) = e^{-B_n t} - \int_0^t F_{\varepsilon_n}(t-u) \left[R_0^{(n)}\left(\frac{u}{\varepsilon_n}\right) V_n \right.$$
$$\left. - V_n B_n \int_0^t R_0^{(n)}\left(\frac{u-v}{\varepsilon_n}\right) e^{-B_n v} dv \right] du. \qquad [3.2]$$

PROOF.– The function $F_{\varepsilon_n}(t)$ satisfies the following perturbed Markov renewal equation

$$F_{\varepsilon_n}(t) = \overline{G}_n\left(\frac{t}{\varepsilon_n}\right) + \int_0^{t/\varepsilon_n} g_n(s)\left(1-\pi_{ke_n}\right)F_{\varepsilon_n}(t-\varepsilon_n s)\,ds, \qquad [3.3]$$

where $\overline{G}_n(t) = 1-G_n(t)$.

Consider the following Laplace transforms:

$$\hat{F}_{\varepsilon_n}(\lambda) = \int_0^\infty e^{-\lambda t}F_{\varepsilon_n}(t)\,dt, \qquad \hat{g}_n(\varepsilon_n\lambda) = \int_0^\infty e^{-\varepsilon_n\lambda t}g_n(t)\,dt,$$

$$\hat{F}_0^{(n)}(\lambda) = \int_0^\infty e^{-\lambda t}F_0^{(n)}(t)\,dt, \quad \hat{R}_0^{(n)}(\varepsilon_n\lambda) = \int_0^\infty e^{-\varepsilon_n\lambda t}R_0^{(n)}(t)\,dt \quad [3.4]$$

$$\hat{Q}^{(n)}(\varepsilon_n\lambda) = \hat{g}_n(\varepsilon_n\lambda)\left(1-\pi_{ke_n}\right), \qquad \hat{\overline{G}}_n(\varepsilon_n\lambda) = \int_0^\infty e^{-\varepsilon_n\lambda t}\overline{G}_n(t)\,dt.$$

Taking into account equations [3.4], then from equation [3.3] it follows that

$$\hat{F}_{\varepsilon_n}(\lambda) = \varepsilon_n\left(1-\hat{Q}^{(n)}(\varepsilon_n\lambda)\right)^{-1}\hat{\overline{G}}_n(\varepsilon_n\lambda).$$

Let us define $S^{(n)}(\varepsilon_n\lambda)$ by the equality

$$\varepsilon_n S^{(n)}(\varepsilon_n\lambda) = m_1^{(n)}\hat{\overline{G}}_n^{-1}(\varepsilon_n\lambda) - 1, \qquad [3.5]$$

i.e. $\hat{\overline{G}}_n(\varepsilon_n\lambda) = m_1^{(n)}\left(1+\varepsilon_n S^{(n)}(\varepsilon_n\lambda)\right)^{-1}$.

It is easily seen that

$$\hat{\overline{G}}_n(\varepsilon_n\lambda) = \int_0^\infty e^{-\varepsilon_n\lambda t}\overline{G}_n(t)\,dt$$

$$= \frac{1}{\varepsilon_n\lambda} - \frac{1}{\varepsilon_n\lambda}\int_0^\infty e^{-\varepsilon_n\lambda t}g_n(t)\,dt = \frac{1}{\varepsilon_n\lambda}\left(1-\hat{g}_n(\varepsilon_n\lambda)\right).$$

Whence, taking into account equations [3.4]–[3.5], we have

$$\hat{Q}^{(n)}(\varepsilon_n\lambda) = \hat{g}_n(\varepsilon_n\lambda)\left(1-\pi_{ke_n}\right)$$

$$= \left(1-\varepsilon_n\lambda m_1^{(n)}\left(1+\varepsilon_n S^{(n)}(\varepsilon_n\lambda)\right)^{-1}\right)\left(1-\pi_{ke_n}\right).$$

Therefore,

$$\hat{F}_{\varepsilon_n}(\lambda) = \varepsilon_n\left(1-\hat{Q}^{(n)}(\varepsilon_n\lambda)\right)^{-1}\hat{\overline{G}}_n(\varepsilon_n\lambda)$$

$$= \varepsilon_n \left(\frac{1+\varepsilon_n S^{(n)}(\varepsilon_n \lambda)}{m_1^{(n)}} - \frac{1+\varepsilon_n \left(S^{(n)}(\varepsilon_n \lambda) - \lambda m_1^{(n)} \right)}{m_1^{(n)}} (1-\pi_{k\varepsilon_n}) \right)^{-1}$$

$$= \left(\lambda + B_n + \varepsilon_n \left(S^{(n)}(\varepsilon_n \lambda) - \lambda m_1^{(n)} \right) B_n \right)^{-1} \qquad [3.6]$$

From equation [3.5], it follows that

$$\varepsilon_n S^{(n)}(\varepsilon_n \lambda) = \frac{\varepsilon_n \lambda m_1^{(n)}}{1 - \hat{g}_n(\varepsilon_n \lambda)} - 1$$

$$= \varepsilon_n \lambda m_1^{(n)} \left(1+\hat{R}_0^{(n)}(\varepsilon_n \lambda) + \frac{1}{\varepsilon_n \lambda m_1^{(n)}} \right) - 1$$

$$= \varepsilon_n \lambda m_1^{(n)} + \varepsilon_n \lambda m_1^{(n)} \hat{R}_0^{(n)}(\varepsilon_n \lambda). \qquad [3.7]$$

Combining equations [3.6]–[3.7], we get

$$\hat{F}_{\varepsilon_n}(\lambda) = \left(\lambda + B_n + \varepsilon_n \lambda m_1^{(n)} B_n \hat{R}_0^{(n)}(\varepsilon_n \lambda) \right)^{-1}.$$

As $\hat{F}_0^{(n)}(\lambda) = \frac{1}{\lambda + B_n}$, then

$$\hat{F}_{\varepsilon_n}^{-1}(\lambda) \hat{F}_0^{(n)}(\lambda) = 1 + \varepsilon_n \lambda m_1^{(n)} B_n \hat{R}_0^{(n)}(\varepsilon_n \lambda) \hat{F}_0^{(n)}(\lambda).$$

Taking into account

$$\frac{1}{\lambda + B_n} = \frac{1}{\lambda} - \frac{B_n}{\lambda} \frac{1}{\lambda + B_n},$$

we have

$$\hat{F}_{\varepsilon_n}(\lambda) = \hat{F}_0^{(n)}(\lambda) - \hat{F}_{\varepsilon_n}(\lambda) \varepsilon_n m_1^{(n)} \hat{R}_0^{(n)}(\varepsilon_n \lambda) B_n \left(1 - B_n \hat{F}_0^{(n)}(\lambda) \right) \qquad [3.8]$$

By taking the inverse Laplace transform of equation [3.8], we conclude the validity of lemma 3.2.

By a component of a distribution $F(x)$, $x \in \mathbb{R}$ we understand a non-negative measure M with the property $0 \neq M \leq F$.

DEFINITION 3.1.– (Sevastyanov 1971). *A distribution function $F(x)$ is said to be spread-out if there exists $k \geq 1$ such that the k-fold convolution $F^{*k}(x)$ has a component M that is absolutely continuous (i.e. it has a density m).*

We will also need the following theorem :

THEOREM 3.1.– (Sevastyanov 1971). If the function $F(x)$ is spread-out and there exist $m_i = \int_0^\infty t^i dF(t)$, $i = 1, 2$, then,

$$\int_0^\infty \left| dH(t) - \frac{dt}{m_1} \right| < \infty,$$

where $H(t) = \sum_{n=1}^\infty F_n(t)$, $F_1(t) = F(t)$, $F_{n+1}(t) = \int_0^t F_n(t-u)\, dF(t)$.

Let us find the asymptotic expansion for $F_{\varepsilon_n}(t) = P\left\{\varepsilon_n \tau_n^{(k)} > t\right\}$ up to $O\left(\varepsilon_n^3\right)$.

THEOREM 3.2.– Let the conditions C1–C3 be fulfilled. In addition, the following condition holds:

C4. $\forall n > k$ $G_n(t)$ is spread-out and there exist all the moments

$$m_2^{(n)} = \int_0^\infty t^2 g_n(t)\, dt.$$

Then for $t \in [\delta, T]$, $0 < \delta < T < \infty$,

$$\begin{aligned}
F_{\varepsilon_n}(t) &= e^{-B_n t} - V_n \int_0^t e^{-B_n(t-u)} R_0^{(n)}\left(\frac{u}{\varepsilon_n}\right) du \\
&+ V_n B_n \int_0^t \int_0^u e^{-B_n(t-(u-y))} R_0^{(n)}\left(\frac{u-y}{\varepsilon_n}\right) dy du \\
&+ V_n^2 \int_0^t \int_0^{t-u} e^{-B_n(t-u-v)} R_0^{(n)}\left(\frac{u}{\varepsilon_n}\right) R_0^{(n)}\left(\frac{v}{\varepsilon_n}\right) dv du \\
&- V_n^2 B_n \int_0^t \int_0^{t-u} e^{-B_n(t-u-v)} R_0^{(n)}\left(\frac{v}{\varepsilon_n}\right) \int_0^u R_0^{(n)}\left(\frac{u-y}{\varepsilon_n}\right) e^{-B_n y} dy dv du \\
&- V_n^2 B_n^2 \int_0^t \int_0^{t-u} e^{-B_n(t-u-v)} \int_0^v R_0^{(n)}\left(\frac{v-y}{\varepsilon_n}\right) e^{-B_n y} dy dv \\
&\times \int_0^u R_0^{(n)}\left(\frac{u-z}{\varepsilon_n}\right) dz du \\
&- V_n^2 B_n \int_0^t \int_0^{t-u} e^{-B_n(t-u-v)} \int_0^v R_0^{(n)}\left(\frac{v-y}{\varepsilon_n}\right) e^{-B_n y} dy dv R_0^{(n)}\left(\frac{u}{\varepsilon_n}\right) du \\
&+ O\left(\varepsilon_n^3\right). \qquad [3.9]
\end{aligned}$$

PROOF.– From theorem 3.1 and the condition C4 it follows that

$$\int_0^\infty |R_0^{(n)}(s)|ds < \infty.$$

After two iterations of equation [3.2], we have

$$F_{\varepsilon_n}(t) = e^{-B_n t} - \int_0^t \left\{ \left[e^{-B_n(t-u)} - \int_0^{t-u} \left\{ e^{-B_n(t-u-v)} \right.\right.\right.$$

$$- \int_0^{t-u-v} F_{\varepsilon_n}(t-u-v-z)$$

$$\times \left(R_0^{(n)}\left(\frac{z}{\varepsilon_n}\right) V_n - V_n B_n \int_0^z R_0^{(n)}\left(\frac{z-x}{\varepsilon_n}\right) e^{-B_n x} dx \right) dz \Bigg\}$$

$$\times \left(R_0^{(n)}\left(\frac{v}{\varepsilon_n}\right) V_n - V_n B_n \int_0^v R_0^{(n)}\left(\frac{v-y}{\varepsilon_n}\right) e^{-B_n y} dy \right) dv \Bigg]$$

$$\left.\times \left(R_0^{(n)}\left(\frac{u}{\varepsilon_n}\right) V_n - V_n B_n \int_0^u R_0^{(n)}\left(\frac{u-y}{\varepsilon_n}\right) e^{-B_n y} dy \right) \right\} du \qquad [3.10]$$

After opening the parentheses in equation [3.10], it is easy to see that all terms, except those contained under integral function $F_{\varepsilon_n}(t)$, are included in equation [3.9].

Assuming conditions C2, C4 and $F_{\varepsilon_n}(t) \leq 1$, we obtain

$$\sup_{n\geq 1} \frac{1}{\varepsilon_n^3} \left| V_n^3 \int_0^t \int_0^{t-u} \int_0^{t-u-v} F_{\varepsilon_n}(t-u-v-z) \right.$$

$$\left. \times R_0^{(n)}\left(\frac{z}{\varepsilon_n}\right) R_0^{(n)}\left(\frac{v}{\varepsilon_n}\right) R_0^{(n)}\left(\frac{u}{\varepsilon_n}\right) dudzdv \right|$$

$$\leq \sup_{n\geq 1} \frac{V_n^3}{\varepsilon_n^3} \left(\int_0^t \left| R_0^{(n)}\left(\frac{s}{\varepsilon_n}\right) \right| ds \right)^3 \leq \left(\int_0^{+\infty} \left| R_0^{(n)}(s) \right| ds \right)^3 < \infty.$$

Here, we are using the following fact:

$$V_n = \frac{\rho_k \pi_k e_n}{\varepsilon_n} \leq 1.$$

Similarly to the arguments for the other terms in equation [3.10], which contain $F_{\varepsilon_n}(t)$, we can find that they are also of the order $O\left(\varepsilon_n^3\right)$. Hence, theorem 3.2 is proved.

Now, let us state the following condition:

C5. There exist the moments for all $n > k$

$$m_3^{(n)} = \int_0^\infty t^3 g_n(t)\, dt.$$

LEMMA 3.3.– Let the conditions C1–C5 hold. Then for all $n > k$

$$\int_0^\infty R_0^{(n)}(t)\,dt = \frac{m_2^{(n)} - 2\left(m_1^{(n)}\right)^2}{2\left(m_1^{(n)}\right)^2} \qquad [3.11]$$

and

$$\int_0^\infty tR_0^{(n)}(t)\,dt = \frac{2m_1^{(n)}m_3^{(n)} - 3\left(m_2^{(n)}\right)^2}{12\left(m_1^{(n)}\right)^3}. \qquad [3.12]$$

PROOF.– Define $\hat{g}_n(s) = \int_0^\infty e^{-st} g_n(t)\,dt$. From conditions C1–C5, it follows that:

$$\int_0^\infty R_0^{(n)}(t)\,dt = \lim_{s\to 0}\int_0^\infty e^{-st}R_0^{(n)}(t)\,dt$$

$$= \lim_{s\to 0}\left(\frac{\hat{g}_n(s)}{1-\hat{g}_n(s)} - \frac{1}{sm_1^{(n)}}\right) = \frac{m_2^{(n)} - 2\left(m_1^{(n)}\right)^2}{2\left(m_1^{(n)}\right)^2}.$$

Similarly, we can prove equation [3.12].

To simplify expressions involving $m_i^{(n)}$, the superscript n in $m_i^{(n)}$ will be omitted, $i = 1, 2, 3$.

THEOREM 3.3.– Let conditions C1–C5 be satisfied. Then, uniformly with respect to $t \in [\delta, T]$, $0 < \delta < T < \infty$

$$P\left\{\varepsilon_n\tau_n^{(k)}>t\right\} = e^{-B_n t} - \varepsilon_n V_n (tB_n - 1)\left(\frac{m_2}{2m_1^2} - 1\right)e^{-B_n t}$$

$$+\varepsilon_n V_n e^{-B_n t}\int_{t/\varepsilon_n}^\infty R_0^{(n)}(s)\,ds$$

$$-\varepsilon_n V_n B_n e^{-B_n t}\int_0^t\int_{u/\varepsilon_n}^\infty R_0^{(n)}(s)\,dsdu - \varepsilon_n^2 B_n V_n e^{-B_n t}\frac{2m_1m_3 - 3m_2^2}{12m_1^3}$$

$$+\varepsilon_n^2 e^{-B_n t}\left[V_n B_n^2 t\frac{2m_1m_3 - 3m_2^2}{12m_1^3} + V_n^2\left(\frac{m_2}{2m_1^2} - 1\right)^2\right]$$

$$-\varepsilon_n^2 V_n^2\left(\frac{m_2}{2m_1^2} - 1\right)^2\left(3 - 3e^{-B_n t} - 2B_n e^{-B_n t} + B_n t e^{-B_n t}\right) + o\left(\varepsilon_n^2\right).$$

Note that the terms

$$\varepsilon_n V_n B_n e^{-B_n t} \int_0^t \int_{u/\varepsilon_n}^{\infty} R_0^{(n)}(s)\, ds du$$

and

$$\varepsilon_n V_n e^{-B_n t} \int_{t/\varepsilon_n}^{\infty} R_0^{(n)}(s)\, ds$$

of this expansion are called boundary layer terms (Korolyuk and Turbin 1982).

PROOF.– Let us analyze the terms on the right side of equation [3.9]. From equations [3.11]–[3.12] and condition C5, it follows that

$$\int_0^t e^{-B_n(t-u)} R_0^{(n)}\left(\frac{u}{\varepsilon_n}\right) du = \varepsilon_n \int_0^{t/\varepsilon_n} \left(e^{-B_n(t-\varepsilon_n s)} - e^{-B_n t}\right) R_0^{(n)}(s)\, ds$$

$$+\varepsilon_n \int_0^{t/\varepsilon_n} e^{-B_n t} R_0^{(n)}(s)\, ds$$

$$= \varepsilon_n^2 B_n e^{-B_n t} \int_0^{t/\varepsilon_n} s R_0^{(n)}(s)\, ds + \varepsilon_n e^{-B_n t} \int_0^{t/\varepsilon_n} R_0^{(n)}(s)\, ds + O\left(\varepsilon_n^3\right)$$

$$= \varepsilon_n^2 B_n e^{-B_n t} \frac{2m_1 m_3 - 3m_2^2}{12 m_1^3} - \varepsilon_n^2 B_n e^{-B_n t} \int_{t/\varepsilon_n}^{\infty} s R_0^{(n)}(s)\, ds$$

$$+\varepsilon_n \left(\frac{m_2}{2m_1^2} - 1\right) e^{-B_n t} - \varepsilon_n e^{-B_n t} \int_{t/\varepsilon_n}^{\infty} R_0^{(n)}(s)\, ds + O\left(\varepsilon_n^3\right)$$

$$= \varepsilon_n \left(\frac{m_2}{2m_1^2} - 1\right) e^{-B_n t} + \varepsilon_n^2 B_n e^{-B_n t} \frac{2m_1 m_3 - 3m_2^2}{12 m_1^3}$$

$$-\varepsilon_n e^{-B_n t} \int_{t/\varepsilon_n}^{\infty} R_0^{(n)}(s)\, ds + o\left(\varepsilon_n^2\right). \qquad [3.13]$$

Similarly, for the following integral, we have

$$\int_0^t \int_0^u e^{-B_n(t-(u-y))} R_0^n \left(\frac{u-y}{\varepsilon_n}\right) dy du$$

$$= \int_0^t e^{-B_n(t-u)} \int_0^u e^{-B_n(u-v)} R_0^n \left(\frac{v}{\varepsilon_n}\right) dv du$$

$$= \int_0^t e^{-B_n(t-u)} \left[\varepsilon_n^2 B_n e^{-B_n u} \frac{2m_1 m_3 - 3m_2^2}{12 m_1^3} - \varepsilon_n^2 B_n e^{-B_n u} \int_{u/\varepsilon_n}^{\infty} s R_0^{(n)}(s)\, ds\right.$$

$$+ \varepsilon_n \left(\frac{m_2}{2m_1^2} - 1 \right) e^{-B_n u} - \varepsilon_n e^{-B_n u} \int_{u/\varepsilon_n}^{\infty} R_0^{(n)}(s) ds + O\left(\varepsilon_n^3\right) \Bigg] du$$

$$= \varepsilon_n^2 B_n e^{-B_n t} t \frac{2m_1 m_3 - 3m_2^2}{12m_1^3} + \varepsilon_n t \left(\frac{m_2}{2m_1^2} - 1 \right) e^{-B_n t}$$

$$+ \varepsilon_n e^{-B_n t} \int_0^t \int_{u/\varepsilon_n}^{\infty} R_0^{(n)}(s)\, ds du + o\left(\varepsilon_n^2\right). \qquad [3.14]$$

Let us show that $\int_0^{t/\varepsilon_n} e^{-B_n(t-\varepsilon_n z)} \int_{t/\varepsilon_n - z}^{\infty} R_0^{(n)}(s)\, ds R_0^{(n)}(z)\, dz \to 0$, as $\varepsilon_n \to 0$.

Indeed,

$$\left| \int_0^{t/\varepsilon_n} e^{-B_n(t-\varepsilon_n z)} \int_{t/\varepsilon_n - z}^{\infty} R_0^{(n)}(s)\, ds R_0^{(n)}(z)\, dz \right|$$
$$\leq \int_0^{t/\varepsilon_n} \int_{t/\varepsilon_n - z}^{\infty} \left| R_0^{(n)}(s) \right| ds \left| R_0^{(n)}(z) \right| dz.$$

Since $\int_0^{\infty} \left| R_0^{(n)}(z) \right| dz$ converges we have $\forall\, \varepsilon > 0$ and there exists $T > 0$, such that $\int_T^{\infty} \left| R_0^{(n)}(z) \right| dz < \varepsilon$. There also exists $L > 0$, such that for a sufficiently small ε_n, $t/\varepsilon_n - L > T$ and $\int_L^{\infty} \left| R_0^{(n)}(z) \right| dz < \varepsilon$.

Thus,

$$\int_0^{t/\varepsilon_n} \int_{t/\varepsilon_n - z}^{\infty} |R_0^{(n)}(s)| ds |R_0^{(n)}(z)| dz = \int_0^L \int_{t/\varepsilon_n - z}^{\infty} |R_0^{(n)}(s)| ds |R_0^{(n)}(z)| dz$$

$$+ \int_L^{t/\varepsilon_n} \int_{t/\varepsilon_n - z}^{\infty} \left| R_0^{(n)}(s) \right| ds \left| R_0^{(n)}(z) \right| dz \leq \varepsilon_n 2 \int_0^{\infty} \left| R_0^{(n)}(s) \right| ds.$$

Therefore, taking onto account equation [3.13], we get

$$\int_0^t \int_0^{t-u} e^{-B_n(t-u-v)} R_0^n \left(\frac{v}{\varepsilon_n} \right) R_0^n \left(\frac{u}{\varepsilon_n} \right) dv du$$

$$= \varepsilon_n \int_0^t \left[\left(\frac{m_2}{2m_1^2} - \frac{1}{2} \right) e^{-B_n(t-u)} + \varepsilon_n B_n e^{-B_n(t-u)} \frac{2m_1 m_3 - 3m_2^2}{12m_1^3} \right.$$

$$\left. - e^{-B_n(t-u)} \int_{(t-u)/\varepsilon_n}^{\infty} R_0^{(n)}(s)\, ds \right] R_0^{(n)} \left(\frac{u}{\varepsilon_n} \right) du + o\left(\varepsilon_n^2\right)$$

$$= \varepsilon_n^2 \left[e^{-B_n t} \left(\left(\frac{m_2}{2m_1^2} - 1 \right)^2 - \left(\frac{m_2}{2m_1^2} - 1 \right) \int_{t/\varepsilon_n}^{\infty} R_0^{(n)}(s)\, ds \right) \right.$$

$$\left. + \int_0^{t/\varepsilon_n} e^{-B_n(t-\varepsilon_n z)} \int_{t/\varepsilon_n - z}^{\infty} R_0^{(n)}(s)\, ds R_0^{(n)}(z)\, dz \right] + o\left(\varepsilon_n^2\right)$$

$$= \varepsilon_n^2 e^{-B_n t} \left(\frac{m_2}{2m_1^2} - 1 \right)^2 + o\left(\varepsilon_n^2\right). \qquad [3.15]$$

For the following term, we have

$$V_n^2 B_n \int_0^t \int_0^{t-u} e^{-B_n(t-u-v)} R_0^n\left(\frac{v}{\varepsilon_n}\right) \int_0^u R_0^n\left(\frac{u-y}{\varepsilon_n}\right) e^{-B_n y} dy dv du$$

$$= V_n^2 B_n \int_0^t \int_0^{t-u} e^{-B_n(t-u-v)} R_0^n\left(\frac{v}{\varepsilon_n}\right) \int_0^u e^{-B_n(u-z)} R_0^n\left(\frac{z}{\varepsilon_n}\right) dz dv du$$

$$= V_n^2 B_n \int_0^t e^{-B_n u} \int_0^{t-u} e^{-B_n(t-u-v)} R_0^n\left(\frac{v}{\varepsilon_n}\right) dv$$

$$\times \left[\varepsilon_n \left(\frac{m_2}{2m_1^2} - 1 \right) \varepsilon_n^2 B_n \frac{2m_1 m_3 - 3m_2^2}{12 m_1^3} - \varepsilon_n \int_{u/\varepsilon_n}^{\infty} R_0^{(n)}(s) ds + o\left(\varepsilon_n^2\right) \right] du$$

$$= \varepsilon_n^2 V_n^2 \left(\frac{m_2}{2m_1^2} - 1 \right)^2 \left(1 - e^{-B_n t}\right) + o\left(\varepsilon_n^2\right). \qquad [3.16]$$

Then,

$$V_n^2 B_n^2 \int_0^t \int_0^{t-u} e^{-B_n(t-u-v)} \int_0^v R_0^n\left(\frac{v-y}{\varepsilon_n}\right) e^{-B_n y} dy dv \int_0^u R_0^n\left(\frac{u-z}{\varepsilon_n}\right) dz du$$

$$= V_n^2 B_n^2 \int_0^t \int_0^{t-u} e^{-B_n(t-u-v)} \int_0^v e^{-B_n(v-w)} R_0^n\left(\frac{w}{\varepsilon_n}\right) dw dv \int_0^u R_0^n\left(\frac{s}{\varepsilon_n}\right) ds du$$

$$= V_n^2 B_n^2 \int_0^t \int_0^{t-u} e^{-B_n(t-u)} \left[\varepsilon_n^2 B_n \frac{2m_1 m_3 - 3m_2^2}{12 m_1^3} + \varepsilon_n \left(\frac{m_2}{2m_1^2} - 1 \right) \right.$$

$$\left. - \varepsilon_n \int_{v/\varepsilon_n}^{\infty} R_0^{(n)}(s)\, ds + o\left(\varepsilon_n^2\right) \right] dv \varepsilon_n \int_0^{u/\varepsilon_n} R_0^n(p)\, dp du$$

$$= \varepsilon_n^2 V_n^2 B_n^2 \left(\frac{m_2}{2m_1^2} - 1 \right)^2 \int_0^t (t-u)\, e^{-B_n(t-u)} du + o\left(\varepsilon_n^2\right)$$

$$= \varepsilon_n^2 V_n^2 \left(\frac{m_2}{2m_1^2} - 1 \right)^2 \left(1 - e^{-B_n t} - B_n e^{-B_n t}\right) + o\left(\varepsilon_n^2\right). \qquad [3.17]$$

Finally,

$$V_n^2 B_n \int_0^t \int_0^{t-u} e^{-B_n(t-u-v)} \int_0^v R_0^{(n)} \left(\frac{v-y}{\varepsilon_n} \right) e^{-B_n y} dy dv R_0^{(n)} \left(\frac{u}{\varepsilon_n} \right) du$$

$$= V_n^2 B_n \int_0^t \int_0^{t-u} e^{-B_n(t-u-v)} \int_0^v e^{-B_n(v-s)} R_0^{(n)} \left(\frac{s}{\varepsilon_n} \right) ds dv R_0^{(n)} \left(\frac{u}{\varepsilon_n} \right) du$$

$$= V_n^2 B_n \int_0^t \int_0^{t-u} e^{-B_n(t-u)} \left(\varepsilon_n^2 B_n \frac{2m_1 m_3 - 3m_2^2}{12 m_1^3} - \varepsilon_n^2 B_n \int_{v/\varepsilon_n}^{\infty} s R_0^{(n)}(s)\, ds \right.$$

$$\left. + \varepsilon_n \left(\frac{m_2}{2m_1^2} - \frac{1}{2} \right) - \varepsilon_n \int_{v/\varepsilon_n}^{\infty} R_0^{(n)}(s)\, ds + O\left(\varepsilon_n^3\right) \right) dv R_0^{(n)} \left(\frac{u}{\varepsilon_n} \right) du$$

$$= \varepsilon_n^2 V_n^2 B_n t \left(\frac{m_2}{2m_1^2} - 1 \right)^2 + o\left(\varepsilon_n^2\right). \qquad [3.18]$$

By adding equations [3.13]–[3.18], we conclude the proof of theorem 3.3.

3.2. Asymptotic inequalities for the distribution of the occupation time of a semi-Markov process in an increasing set of states

We will assume that condition C2 of section 3.1 is not fulfilled. We then establish estimates for the distribution of the first hitting time of an infinitely remote level (or hard-to-reach state) of a semi-Markov process. We will number the states on the semi-axis of positive integer numbers. Since $\forall\, n > k$

$$0 \leq \frac{\rho_k \pi_{k e_n}}{\varepsilon_n} \leq 1,$$

then, there exist the limits

$$\underline{\lim}_{n \to \infty} \frac{\rho_k \pi_{k e_n}}{\varepsilon_n} = A_0, \qquad \overline{\lim}_{n \to \infty} \frac{\rho_k \pi_{k e_n}}{\varepsilon_n} = A_1. \qquad [3.19]$$

It was proved in Korolyuk *et al.* (1979b) that if the following limit exists:

$$\lim_{n \to \infty} \frac{\rho_k \pi_{k e_n}}{\varepsilon_n} = A, \qquad [3.20]$$

then A does not depend on k. Similarly, we can show that A_0, A_1 do not depend on k.

THEOREM 3.4.– Let conditions C1, C3, C4 and C5 of theorem 3.1 be satisfied. Furthermore, suppose that

$$\lim_{n \to \infty} m_2^{(n)} = m_2 < \infty.$$

Then, uniformly with respect to $t \in [\sigma, T], 0 < \sigma < T$

$$\exp\left\{-t \sup_{i \geq n} B_i\right\} + \varepsilon_n \min_{\lambda \in [\lambda_0, \lambda_1]} f(\lambda, t, m_1, m_2) + o(\varepsilon_n) \leq P\left\{\varepsilon_n \tau_n^{(k)} \geq t\right\}$$

$$\leq \exp\left\{-t \inf_{i \geq n} B_i\right\} + \varepsilon_n \max_{\lambda \in [\lambda_0, \lambda_1]} f(\lambda, t, m_1, m_2) + o(\varepsilon_n), \qquad [3.21]$$

where

$$\lambda_0 = \frac{A_0}{m_1}, \qquad \lambda_1 = \frac{A_1}{m_1}, \qquad m_1 = \sum_{i=1}^{\infty} \rho_i a_i,$$

$$f(\lambda, t, m_1, m_2) = \left(\lambda^2 t - \lambda\right)\left(m_1 - m_2/2m_1\right) e^{-\lambda t}.$$

PROOF.– By iterating equation [3.2], we obtain

$$F_{\varepsilon_n}(t) = e^{-B_n t} - \int_0^t \left\{e^{-B_n(t-u)} - \int_0^{t-u} F_{\varepsilon_n}(t-u-y)\left[R_0^{(n)}\left(\frac{y}{\varepsilon_n}\right) V_n\right.\right.$$

$$\left.\left. -V_n B_n \int_0^y R_0^{(n)}\left((y-s)/\varepsilon_n\right) e^{-B_n s} ds\right] dy\right\} \left[R_0^{(n)}\left(\frac{u}{\varepsilon_n}\right) V_n\right.$$

$$\left. -V_n B_n \int_0^u R_0^{(n)}\left(\frac{u-v}{\varepsilon_n}\right) e^{-B_n v} dv\right] du. \qquad [3.22]$$

Let us analyze the terms in the right-hand side of equation [3.22]. Taking into account condition C4, we have

$$\int_0^t e^{-B_n(t-u)} R_0^{(n)}\left(\frac{u}{\varepsilon_n}\right) V_n \, du$$

$$= \varepsilon_n V_n \int_0^{t/\varepsilon_n} \left(e^{-B_n(t-\varepsilon_n s)} - e^{-B_n t}\right) R_0^{(n)}(s)\, ds$$

$$+ \varepsilon_n V_n e^{-B_n t} \int_0^{t/\varepsilon_n} R_0^{(n)}(s)\, ds$$

$$= \varepsilon_n \left(m_1^{(n)}\right)^{-2} \left(m_2^{(n)}/2 - \left(m_1^{(n)}\right)^2\right) V_n e^{-B_n t} + o(\varepsilon_n), \qquad [3.23]$$

where we have used the result

$$\int_0^\infty R_0^{(n)}(s)\, ds = \left(m_2^{(n)}/2\left(m_1^{(n)}\right)^2 - 1\right).$$

Similarly,

$$B_nV_n\int_0^t e^{-B_n(t-u)}\int_0^u R_0^{(n)}\left((u-v)/\varepsilon_n\right)e^{-B_nv}dvdu$$

$$=\varepsilon_nB_nV_n\int_0^t e^{-B_n(t-u)}\int_0^{u/\varepsilon_n} R_0^{(n)}(s)\,e^{-B_n(u-\varepsilon_ns)}dsdu$$

$$=\varepsilon_nB_nV_nte^{-B_nv}\left(m_1^{(n)}\right)^{-2}\left(m_2^{(n)}/2-\left(m_1^{(n)}\right)^2\right)+o\left(\varepsilon_n\right). \qquad [3.24]$$

Considering that $F_{\varepsilon_n}(t)\leq 1$, and condition C4, we obtain

$$V_n^2\left|\int_0^t\int_0^{t-u}F_{\varepsilon_n}(t-u-y)\,R_0^{(n)}(y/\varepsilon_n)\,R_0^{(n)}(u/\varepsilon_n)\,dydu\right|$$

$$\leq\varepsilon_n^2\left(\int_0^\infty\left|R_0^{(n)}(s)\right|ds\right)^2=o\left(\varepsilon_n\right). \qquad [3.25]$$

It was proved in Korolyuk *et al.* (1979b) that

$$\lim_{n\to\infty}m_1^{(n)}=m_1=\sum_{i=1}^\infty\rho_i\,a_i. \qquad [3.26]$$

Taking into account equation [3.26] and condition C3, we have

$$\left|B_n^2V_n^2\int_0^t\int_0^{t-u}F_{\varepsilon_n}(t-u-y)\int_0^y R_0^{(n)}\left((y-s)/\varepsilon_n\right)e^{-B_ns}dsdy\right.$$

$$\left.\times\int_0^u R_0^{(n)}\left((u-v)/\varepsilon_n\right)e^{-B_nv}dvdu\right|$$

$$\leq\frac{t}{m_1^2}\varepsilon_n^2\left(\int_0^\infty\left|R_0^{(n)}(s)\right|ds\right)^2=o\left(\varepsilon_n\right). \qquad [3.27]$$

In a similar manner, we can prove that all other terms in the right-hand side of equation [3.22] containing $F_{\varepsilon_n}(t)$ are of the order $o\left(\varepsilon_n\right)$.

Passing to the upper and lower limits in equations [3.23]–[3.24] as $n\to\infty$, and taking into account both equations [3.25]–[3.27] and condition C4, we can now finish the proof of theorem 3.4.

As a result of equation [3.21], we have

$$e^{-\lambda_1t}+o(1)\leq P\left\{\varepsilon_n\tau_n^{(k)}\geq t\right\}\leq e^{-\lambda_0t}+o(1),$$

where $o(1)\to 0$ as $n\to+\infty$.

REMARK 3.2.– *It was proved in Korolyuk* et al. *(1979b) that if*

$$\lim_{n\to\infty} V_n = A > 0 \quad \textit{then} \quad \lim_{n\to\infty} P\left\{\varepsilon_n \tau_n^{(k)} \geq t\right\} = e^{-\lambda t}$$

where $\lambda = A/m_1$. *We have obtained the generalization of this result without considering the requirement of the existence of the limit for A.*

Now, the results of section 3.1, with some additional conditions depending on l, allow us to write inequalities [3.21] up to $o\left(\varepsilon_n^l\right)$, $\forall\, l \geq 1$.

COROLLARY 3.1.– Let us assume that a semi-Markov process $\{\xi(t)\}$ satisfies conditions C1, C3 and C4 of theorem 3.2.

Then, uniformly, with respect to $t \in [\sigma, S]$, $0 < \sigma < S < \infty$,

$$P\left\{\pi_{ke_n}\tau_n^{(k)} \geq t\right\} = e^{-t/m_1^{(n)}}$$

$$+\pi_{ke_n}\left(\frac{t}{m_1^{(n)}} - 1\right)\left(\frac{m_2^{(n)}}{2\left(m_1^{(n)}\right)^2} - 1\right) e^{-t/m_1^{(n)}} + o\left(\pi_{ke_n}\right).$$

It is easily seen that if $m_2 = \lim_{n\to\infty} m_2^{(n)}$, then

$$P\left\{\pi_{ke_n}\tau_n^{(k)} \geq t\right\} = e^{-t/m_1^{(n)}} + \pi_{ke_n}\left(\frac{t}{m_1} - 1\right)\left(\frac{m_2}{2m_1^2} - 1\right) e^{-t/m_1}$$

$$+o\left(\pi_{ke_n}\right),$$

where $m_1 = \sum_{i=1}^{\infty} \rho_i\, a_i$ (formula [3.26]).

3.3. Asymptotic analysis of the occupation time distribution of an embedded semi-Markov process (with increasing states) in a diffusion process

Let $\{\eta(t), t \geq 0\}$ be a diffusion process with the infinitesimal operator

$$\mathfrak{A} = \frac{\sigma^2(x)}{2}\frac{d^2}{dx^2} + a(x)\frac{d}{dx}, \qquad [3.28]$$

where $\sigma^2(x)$, $a(x)$ are continuous functions on $\mathbb{R}$. Let us state the following condition:

A1. *Suppose*

$$\int_{-\infty}^{\infty} \left(\sigma^2(x)\right)^{-1} e^{\int_{-\infty}^{x} 2a(u)/\sigma^2(u)du} dx < \infty.$$

It is well-known (Gikhman and Skorokhod 1975; Feller 1971) that under condition A1 there exists the stationary distribution of the process $\eta(t)$.

Let us define a semi-Markov process embedded in the diffusion process $\eta(t)$. Now, we fix a real number $\Delta > 0$ to define the following partition:

$$X = \{\ldots, -2\Delta, -\Delta, 0, \Delta, 2\Delta, \ldots\}.$$

Assume that $\eta(0) = 0$. Consider a sequence of random variables $\{\tau_j, j \geq 0\}$ as follows: specify $\tau_0 = 0$, and by τ_1 denote the first exit time of the process $\eta(t)$ from $(-\Delta, \Delta)$; next, suppose τ_i is the first exit time of $\eta(t)$ from $((l-1)\Delta, (l+1)\Delta)$, then τ_{i+1} is the first exit time of $\eta(t)$ from $(l\Delta, (l+2)\Delta)$ if $\eta(\tau_i) = (l+1)\Delta$ or the first exit time of $\eta(t)$ from $((l-2)\Delta, l\Delta)$ if $\eta(\tau_i) = (l-1)\Delta$.

Thus, for $i \geq 1$

$$\tau_i = \inf\{t > \tau_{i-1} : \eta(t) \notin [\eta(\tau_{i-1}) - \Delta, \eta(\tau_{i-1}) + \Delta]\}.$$

Since $\eta(t)$ is a diffusion process and therefore, strictly Markov, the sequence $\{\eta(\tau_j), j \geq 0\}$ is a Markov walk on the phase space X.

Define $h(t) = \eta(\tau_j)$ for $\tau_j \leq t < \tau_{j+1}$. It is easy to see that the process $\{h(t)\}$ is semi-Markov with the embedded Markov chain $\{\eta(\tau_j), j \geq 0\}$.

It is also easily verified that if $h(0) = \eta(0)$, then the first reaching time of a level $m\Delta$, $m \in Z$ by the process $h(t)$ coincides with the first reaching time of the same level by the corresponding diffusion processes $\eta(t)$.

In order to apply corollary 3.1 to the semi-Markov process $h(t)$, we define π_{0e_T}, m_1^T, m_2^T, where $T \in X$, $(T = k\Delta)$, $k > 0$, and e_T is the lumped state after merging the set of states

$$X^T = X \backslash X_T,$$

where

$$X_T = \{-(k-1)\Delta, -(k-2)\Delta, \cdots, -\Delta, 0, \Delta, \cdots, (k-2)\Delta, (k-1)\Delta\},$$

m_1^T, m_2^T are the first and the second moments, respectively, for the hitting time (or reaching time) of the set X^T by the process $\eta(t)$.

Denote by $f_{0,\Delta}$ (respectively, $f_{0,-\Delta}$) the probability of a hitting time of Δ (respectively, $-\Delta$) by the diffusion process $\eta(t)$, which started from 0 and did not reach $-\Delta$ (respectively, Δ).

The function $f_{0,\Delta x}$ is a solution for the following equation:

$$\frac{\sigma^2(x)}{2}\frac{d^2 f}{dx^2} + a(x)\frac{df}{dx} = 0 \tag{3.29}$$

with the boundary conditions $f(-\Delta) = 0,\ f(\Delta) = 1$ (Gikhman and Skorokhod 1975).

By solving equation [3.29], we have

$$f_{0,\Delta x} = \frac{\int_{-\Delta}^{0} e^{-\int_{-\Delta}^{y} 2a(u)/(\sigma(u))^2 du} dy}{\int_{-\Delta}^{\Delta} e^{-\int_{-\Delta}^{y} 2a(u)/(\sigma(u))^2 du} dy}. \tag{3.30}$$

Similarly, solving equation [3.29] with the boundary conditions $f(-\Delta) = 1$, $f(\Delta) = 0$, we have

$$f_{0,-\Delta x} = \frac{\int_{0}^{\Delta} e^{-\int_{-\Delta}^{y} 2a(u)/(\sigma(u))^2 du} dy}{\int_{-\Delta}^{\Delta} e^{-\int_{-\Delta}^{y} 2a(u)/(\sigma(u))^2 du} dy}.$$

Denote by $f_{\Delta,T}$ (respectively, $f_{-\Delta,-T}$) the hitting time probability of reaching the set X^T by the diffusion process $\eta(t)$, which started from Δ (respectively, $-\Delta$) and did not reach 0. The function $f_{\Delta,T}$ is a solution of equation [3.29] with the boundary conditions $f(0) = 0,\ f(T) = 1$:

$$f_{\Delta,T} = \frac{\int_{0}^{\Delta} e^{-\int_{0}^{y} 2a(u)/(\sigma(u))^2 du} dy}{\int_{0}^{T} e^{-\int_{0}^{y} 2a(u)/(\sigma(u))^2 du} dy}.$$

Similarly, the function $f_{-\Delta,-T}$ is a solution of equation [3.29] with the boundary conditions $f(0) = 0,\ f(-T) = 1$:

$$f_{-\Delta,-T} = \frac{\int_{-\Delta}^{0} e^{-\int_{-T}^{y} 2a(u)/(\sigma(u))^2 du} dy}{\int_{-T}^{0} e^{-\int_{-T}^{y} 2a(u)/(\sigma(u))^2 du} dy}.$$

Thus,

$$\pi_{0\, e_T} = f_{0,\Delta}\, f_{\Delta,T} + f_{0,-\Delta}\, f_{-\Delta,-T}. \tag{3.31}$$

The expectation m_1^T of the occupation time of the process $h(t)$ in the set X_T coincides with the expectation of the first hitting time of the set X^T by the process $\eta(t)$. This expected value can be obtained from the following equation:

$$\frac{\sigma^2(x)}{2}\frac{d^2m}{dx^2}+a(x)\frac{dm}{dx}=-1, \qquad [3.32]$$

with the boundary conditions $m(-T)=0$, $m(T)=0$ (Gikhman and Skorokhod 1975).

By solving equation [3.32], we have (Gikhman and Skorokhod 1975)

$$m_1^T=m(0)=-2\frac{\varphi(0)}{\varphi(T)}\int_{-T}^{T}\frac{\varphi(T)-\varphi(z)}{\sigma^2(z)\varphi'(z)}dz+2\int_{-T}^{0}\frac{\varphi(0)-\varphi(z)}{\sigma^2(z)\varphi'(z)}dz \quad [3.33]$$

where

$$\varphi(x)=\int_{-T}^{x}\exp\left\{-\int_{-T}^{y}2a(u)/(\sigma(u))^2du\right\}dy.$$

Let m_2^T be the second moment of the hitting time of T by the process $\eta(t)$ provided that $\eta(t)=0$. It is well-known (Gikhman and Skorokhod 1975) that m_2^T is a solution of the following equation:

$$\frac{\sigma^2(x)}{2}\frac{d^2m_2}{dx^2}+a(x)\frac{dm_2}{dx}=-2m_1(x), \qquad [3.34]$$

with boundary conditions $m_2(-T)=0$, $m_2(T)=0$.

By solving equation [3.34], we have

$$m_2^T=m_2(0)=C\frac{\int_{-T}^{0}\exp\left\{-\int_{-T}^{s}2a(u)/(\sigma(u))^2du\right\}ds}{\int_{-T}^{T}\exp\left\{-\int_{-T}^{s}2a(u)/(\sigma(u))^2du\right\}ds} \qquad [3.35]$$

$$-\int_{-T}^{0}\int_{-T}^{s}2m_1(z)\exp\left\{\int_{-T}^{z}2a(u)/(\sigma(u))^2du\right\}dz$$
$$\times\exp\left\{-\int_{-T}^{s}2a(u)/(\sigma(u))^2du\right\}ds,$$

where

$$C=\int_{-T}^{T}\int_{-T}^{s}2m_1(z)\exp\left\{\int_{-T}^{z}2a(u)/(\sigma(u))^2du\right\}dz$$
$$\times\exp\left\{-\int_{-T}^{s}2a(u)/(\sigma(u))^2du\right\}ds.$$

Let ν_0^T be a sojourn time at state 0 of the lumped semi-Markov process $h_T^0(t)$, defined on the phase space $X_0^T = \{0\} \cup e_T$, where e_T is a lumping state after merging the set of states X^T.

Denote by $G_T(t) = P\{\nu_0^T \le t\}$. Then $m_i^T = E(\nu_0^T)^i$, $i = 1, 2, \cdots$.

The moments $m_j^T = m_j(0)$ can be calculated as follows (Gikhman and Skorokhod 1975):

$$\frac{\sigma^2(x)}{2} \frac{d^2 m_j}{dx^2} + a(x) \frac{dm_j}{dx} = -j m_{j-1}(x), \ m_j(-T) = m_j(T) = 0. \quad [3.36]$$

Let us state the following condition:

A2. *For any sufficiently large $T > 0$, the distribution $G_T(t)$ is spread-out and there exists the moment $m_2^T = \int_0^\infty t^2 \, dG_T(t)$.*

Denote by $\theta_T^{(b)}$ the first exit time from X_T of the diffusion process $\eta(t)$ provided $\eta(0) = b \in X$, $|b| < T$, i.e.

$$\theta_T^{(b)} = \inf \{t : |\eta(t)| = T, \ \eta(0) = b\}.$$

THEOREM 3.5.– Suppose conditions A1 and A2 are fulfilled. Then, uniformly with respect to $t \in [\sigma, S]$, $0 < \sigma < S < \infty$,

$$P\left\{\varepsilon_T \theta_T^{(0)} \ge t\right\}$$
$$= e^{-(m_1^T)^{-1} t} + \varepsilon_T \left(t/m_1^T - 1\right) \left(m_2^T/2(m_1^T)^2 - 1\right) e^{-(m_1^T)^{-1} t} + o(\varepsilon_T),$$

where $\varepsilon_T = \pi_{0\, e_T}$ is defined in equation [3.31] and m_1^T, m_2^T can be obtained from equations [3.33]–[3.36].

PROOF.– This result can be proved after applying corollary 3.1 to the semi-Markov process $h(t)$, and taking into account that the hitting time of set X^T by the $h(t)$, provided $h(0) = 0$, is equal in distribution to $\theta_T^{(0)}$.

EXAMPLE 3.2.– Let $\tau_{[-T, T]}$, $T > 0$, be an exit time from $[-T, T]$ of the Ornstein–Uhlenbeck process $\{\xi(t)\}$ with the infinitesimal operator $\mathfrak{A} = \frac{\partial^2}{\partial x^2} - x \frac{\partial}{\partial x}$. The process $\xi(t)$ satisfies conditions A1–A2 and we can apply theorem 3.5 to the random variable $\theta_T^{(0)} = \inf \{t : |\xi(t)| = T, \ \xi(0) = 0\}$.

3.4. Asymptotic analysis of a semigroup of operators of the singularly perturbed random evolution in semi-Markov media

In this section, we will deal with a random evolution $u_\varepsilon(t,x)$ in semi-Markov media $\xi_\varepsilon(t/\varepsilon^2)$ that depends on a small parameter $\varepsilon > 0$. It was proved in Korolyuk (1993), Korolyuk and Swishchuk (1995b) and Korolyuk and Limnios (2004), that under the balance condition for a fixed $t > 0$, the process $u_\varepsilon(t,x)$ weakly converges to the diffusion process.

By using a common method for solutions of singularly perturbed equations, we obtain asymptotic expansions in the form of a series of regular and singular terms for the semigroup of operators of a corresponding three-component Markov process provided that $u_\varepsilon(t,x)$ weakly converges to a diffusion process as $\varepsilon \to 0$.

Let $\{\xi(t),\ t \geq 0\}$ be a semi-Markov process on the phase space $(E,\mathcal{F})$ with the semi-Markov kernel

$$Q(x,B,t) = P(x,B)\,G_x(t),\quad B \in \mathcal{F},$$

where $P(x,B)$ are the transition probabilities of the embedded Markov chain $\{\xi_n,\ n \geq 0\}$, and $G_x(t)$ is the cdf of a sojourn time (holding time) of $\xi(t)$ in $x \in E$.

Now, let us assume a function $C(u,x)$, $u \in \mathbb{R}$, $x \in E$ such that it satisfies the unique valued solvability condition of the following evolution equation:

$$\frac{du(t,x)}{dt} = C(u(t,x),\xi(t)),\qquad u(0,x) = u_0. \qquad [3.37]$$

In addition, we assume that the derivative $\frac{\partial}{\partial u}C(u,x)$ is bounded for all $x \in E$.

For the fixed parameter $\varepsilon > 0$, consider the following random *transport* process $u_\varepsilon(t)$ in the scaled semi-Markov medium $\xi(t/\varepsilon^2)$

$$\frac{du_\varepsilon(t)}{dt} = \frac{1}{\varepsilon}C\left(u_\varepsilon(t),\xi\left(t/\varepsilon^2\right)\right),\qquad u_\varepsilon(0) = u_0.$$

Let us assume that the following four conditions hold:

B1. There exists the pdf $g_x(t) = \frac{dG_x(t)}{dt}$ and the first two moments $m_x^{(1)} = \int_0^\infty t\,g_x(t)\,dt$, $m_x^{(2)} = \int_0^\infty t^2 g_x(t)\,dt$ for all $x \in E$.

B2. The embedded Markov chain ξ_n, $n = 0,1,2,\dots$ is uniformly ergodic (Kartashov 1985) with the stationary distribution $\rho(\cdot)$, i.e.

$$\lim_{n\to\infty}\left\|\frac{1}{n}\sum_{i=1}^{n}P^{(i)} - \Pi_0\right\| = 0,$$

where $P^{(i)}(x,y) = \int_E P^{(i-1)}(x,dz)\,P(z,y)$ with $P(z,y) = P^{(1)}(z,y)$, and $\Pi_0 h(x) = \int_E \rho(ds)\,h(s)\mathbf{1}(x)$ is the projector of the Markov chain ξ_n (Volume 1, Chapter 1, section 1.3) (Korolyuk and Turbin 1993).

B3. The following supremum is bounded:

$$\sup_{x,u} \frac{\int_u^\infty \overline{G}_x(t)dt}{\overline{G}_x(u)} < \infty, \quad \overline{G}_x(t) = 1 - G_x(t).$$

B4. The following moments are positive:

$$\hat{m}^{(k)} = \int_X \rho(dx) m_x^{(k)} > 0, \ \ k = 1, 2.$$

Denote by $\mathbb{X} = E \times [0, \infty)$ and by $\mathcal{X} = \mathcal{F} \times \mathcal{B}_+$, where $\mathcal{B}_+$ is a Borel σ-algebra on $[0, \infty)$.

On the phase space $\mathbb{X}$, we consider the following couple of processes:

$$\{\varsigma(t) = (\xi(t), \tau(t)),\ t \geq 0\},$$

where $\tau(t) = t - \sup\{u \leq t\colon \xi_\varepsilon(u) \neq \xi_\varepsilon(t)\}$.

Denote by $\mathbb{B}$ the Banach space of bounded $\mathcal{X}$-measurable functions on X with supremum norm

$$\|\varphi\| = \sup_{x,\tau} |\varphi(x,\tau)| < \infty.$$

It is well-known that $\varsigma(t)$ is a Markov process and its infinitesimal operator Q can be written as (see Volume 1, Chapter 1, section 1.4):

$$Q\varphi(x,\tau) = r_x(\tau)\left[P\varphi(x,0) - \varphi(x,\tau)\right] + \frac{\partial}{\partial\tau}\varphi(x,\tau),$$

where

$$r_x(t) = \frac{g_x(t)}{1-G_x(t)},$$

$$P\varphi(x,0) = \int_E P(x,dy)\,\varphi(y,0),$$

and $\varphi(x,\tau) \in \mathbb{B}$ is a continuously differentiable function with respect to τ.

Let us introduce the operator $\Pi_1 f = (\pi, f)\mathbb{I}$, where $f \in \mathbb{B}$ and $\mathbb{I}(x,\tau) = 1$, for all $(x,\tau) \in X$,

$$\pi(B \times [0,s]) = \frac{1}{\hat{m}^{(1)}} \int_B \int_0^s \rho(dx)\,\overline{G}_x(s)ds,$$

and the inner product

$$(\pi, g) = \int_X \pi (dz)\, g(z) = \frac{1}{\hat{m}^{(1)}} \int_X \int_0^{\infty} \rho (dx)\, \overline{G}_x(s) g(x, s)\, ds.$$

It is well-known that Π_1 is the projection operator on $N(Q)$ (Volume 1, Chapter 1, section 1.3), (Korolyuk and Turbin 1982), i.e.

$$\Pi_1 Q = Q\Pi_1 = 0.$$

As was mentioned in Volume 1, Chapter 1, section 1.3, let $\mathbf{R}_0$ be the potential operator of the embedded Markov chain ξ_n, i.e. $\mathbf{R}_0 = \Pi_0 - (P + \Pi_0)^{-1}$.

It is proved in lemma 4.2 of Korolyuk and Turbin (1982) that under conditions C1–C4 the potential operator $\mathcal{R}_0$ of $\varsigma(t)$ is given by

$$\begin{aligned} \mathcal{R}_0 f(x, u) &= \int_0^{\infty} \frac{\overline{G}_x(t)}{\overline{G}_x(u)} f(x, t)\, dt - (\pi, f) \frac{\int_0^{\infty} \overline{G}_x(t)\, dt}{\overline{G}_x(u)} \\ &- \frac{1}{\hat{m}^{(1)}} \int_X \int_0^{\infty} \int_y^{\infty} \rho (dx)\, \overline{G}_x(z)\, f(x, z)\, dz dy + \frac{\hat{m}^{(2)}}{2\hat{m}^{(1)}} (\pi, f) \\ &+ (I - \Pi_1)\, P\mathcal{R}_0 \int_0^{\infty} \overline{G}_x(z)\, f(x, z)\, dz - (\pi, f)\, (I - \Pi_1)\, P\mathcal{R}_0 m(x). \end{aligned}$$

We should recall that the potential operator of $\varsigma(t)$ is the generalized inverse operator of Q and it satisfies $\mathcal{R}_0 Q = Q\mathcal{R}_0 = I - \Pi_1$ (Volume 1, Chapter 1, section 1.3).

It is also well-known that the three-component process

$$\varsigma(t) = (u(t, x), \xi(t), \tau(t))$$

is a Markov process on the phase space $\mathbb{R} \times E \times [0, \infty)$ with the infinitesimal operator A given by (Volume 1, Chapter 1, section 1.4), (Korolyuk and Turbin 1982; Corlat *et al.* 1991):

$$\begin{aligned} A\varphi(u, x, \tau) &= \mathbb{C}(u)\, \varphi(u, x, \tau) + Q\varphi(u, x, \tau) \\ &= \mathbb{C}(u)\, \varphi(u, x, \tau) + r_x(\tau)\, [P\varphi(u, x, 0) - \varphi(u, x, \tau)] + \frac{\partial}{\partial \tau}\varphi(u, x, \tau), \end{aligned}$$

where $\varphi(u, z, \tau)$ is a continuously differentiable function with respect to u and τ, and it is in the domain of the operator A. We have

$$\mathbb{C}(u)\, f(u, x) = C(u, x)\, \frac{\partial}{\partial u} f(u, x),$$

$$P\varphi(u,z,0)=\int_E P(z,dy)\,\varphi(u,y,0).$$

Let us consider the scaled process $\varsigma_\varepsilon(t)=\left(u_\varepsilon(t,x),\xi_\varepsilon\left(t/\varepsilon^2\right),\tau\left(t/\varepsilon^2\right)\right)$. It can be easily verified that its infinitesimal operator A satisfies:

$$A_\varepsilon\varphi(u,x,\tau)=\frac{1}{\varepsilon}\mathbb{C}(u)\,\varphi(u,x,\tau)+\frac{1}{\varepsilon^2}r_x(\tau)\left[P\varphi(u,x,0)-\varphi(u,x,\tau)\right]$$
$$+\frac{1}{\varepsilon^2}\frac{\partial}{\partial\tau}\varphi(u,x,\tau).$$

By defining $\varphi_\varepsilon(t,u,x,\tau)=\mathbf{E}\left[\varphi(\varsigma_\varepsilon(t))|\,\varsigma_\varepsilon(0)=(u,x,\tau)\right]$, we have (for notational conciseness we define t,u,x,τ as $\boldsymbol{\theta}$ in the rest of this chapter)

$$\frac{\partial}{\partial t}\varphi_\varepsilon(\boldsymbol{\theta})=A_\varepsilon\varphi_\varepsilon(\boldsymbol{\theta})=\frac{1}{\varepsilon}\mathbb{C}(u)\,\varphi_\varepsilon(\boldsymbol{\theta})$$
$$+\frac{1}{\varepsilon^2}r_x(\tau)\left[P\varphi_\varepsilon(t,u,x,0)-\varphi_\varepsilon(\boldsymbol{\theta})\right]+\frac{1}{\varepsilon^2}\frac{\partial}{\partial\tau}\varphi_\varepsilon(\boldsymbol{\theta}), \qquad [3.38]$$

with the boundary condition $\varphi_\varepsilon(0,u,x,\tau)=\varphi^{(0)}$. We will assume $\varphi^{(0)}(u)\in C^\infty$.

THEOREM 3.6.– Suppose that conditions C1–C4 are fulfilled. In addition, assume that the following balance condition holds:

B5. $\Pi_1\mathbb{C}(u)\,\Pi_1=0$.

Then, the solution of [3.38] can be expanded in the following form:

$$\varphi_\varepsilon(\boldsymbol{\theta})=\varphi^{(0)}(\boldsymbol{\theta})+\sum_{n=1}^{\infty}\varepsilon^n\left(\varphi^{(n)}(\boldsymbol{\theta})+g^{(n)}\left(\frac{t}{\varepsilon^2},u,x,\tau\right)\right),$$

where the first regular term $\varphi^{(0)}(\boldsymbol{\theta})$ satisfies the diffusion equation

$$\frac{\partial}{\partial t}\varphi^{(0)}(\boldsymbol{\theta})+\Pi_1 C(u,x)\,\frac{\partial}{\partial u}\mathcal{R}_0 C(u,x)\,\frac{\partial}{\partial u}\Pi_1\varphi^{(0)}(\boldsymbol{\theta})=0,$$

and all the terms $\varphi^{(k)}(\boldsymbol{\theta})$, $k\geq 1$ can be calculated in a recursive manner. The first singular term is of the following form:

$$g^{(1)}(\boldsymbol{\theta})=g^{(1)}(0,u,x,\tau)\exp_0\{Qt\},$$

where $\exp_0\{Qt\}=\exp\{Qt\}-\Pi_1$ and all the terms $k\geq 2$ can be recursively calculated in the following form:

$$g^{(k+1)}(\boldsymbol{\theta})=\exp_0\{Qt\}\,g^{(k+1)}(0,u,x,\tau)$$
$$+\int_0^t \exp_0\{Q(t-s)\}C(u,x)\,\frac{\partial}{\partial u}g^{(k)}(\boldsymbol{\theta})\,ds.$$

PROOF.– By applying the method for singularly perturbed equations developed by Korolyuk and considered in Korolyuk and Turbin (1982), we can find a solution of [3.38] in the following form:

$$\varphi_\varepsilon(\boldsymbol{\theta}) = \varphi^{(0)}(\boldsymbol{\theta}) + \sum_{n=1}^{\infty} \varepsilon^n \Big(\varphi^{(n)}(\boldsymbol{\theta}) + g^{(n)}\Big(\frac{t}{\varepsilon^2}, u, x, \tau\Big)\Big)$$

and

$$\varphi^{(0)}(0, u, x, \tau) + \sum_{n=1}^{\infty} \varepsilon^n \left(\varphi^{(n)}(0, u, x, \tau) + g^{(n)}(0, u, x, \tau)\right) = \varphi_\varepsilon^{(0)} \quad [3.39]$$

where $\varphi^{(n)}(\boldsymbol{\theta})$, $n \geq 0$, are regular terms and $g^{(n)}\left(\frac{t}{\varepsilon^2}, u, x, \tau\right)$, $n \geq 1$, are singular terms of expansion [3.39].

Substituting [3.39] into [3.38], we have for the regular term $\varphi^{(0)}(\boldsymbol{\theta})$:

$$Q\varphi^{(0)}(\boldsymbol{\theta}) = r_x(\tau)\left[P\varphi^{(0)}(t, u, x, 0) - \varphi^{(0)}(\boldsymbol{\theta})\right] + \frac{\partial}{\partial \tau}\varphi^{(0)}(\boldsymbol{\theta}) = 0.$$

Therefore, $\varphi^{(0)}(\boldsymbol{\theta}) \in ker(Q)$. Then, we can obtain for $\varphi^{(1)}(\boldsymbol{\theta})$,

$$C(u, x)\frac{\partial}{\partial u}\varphi^{(0)}(\boldsymbol{\theta}) + r_x(\tau)\left[P\varphi^{(1)}(t, u, x, 0) - \varphi^{(1)}(\boldsymbol{\theta})\right] + \frac{\partial}{\partial \tau}\varphi^{(1)}(\boldsymbol{\theta}) = 0. \quad [3.40]$$

Hence,

$$\varphi^{(1)}(\boldsymbol{\theta}) = -\mathcal{R}_0 C(u, x)\frac{\partial}{\partial u}\varphi^{(0)}(\boldsymbol{\theta}) + n_1(\boldsymbol{\theta}), \quad [3.41]$$

where $n_1(\boldsymbol{\theta}) \in ker(Q)$, and it depends on the initial conditions of equation [3.39].

Now, for $k \geq 2$ we have

$$\frac{\partial}{\partial t}\varphi^{(k-2)}(\boldsymbol{\theta}) = C(u, x)\frac{\partial}{\partial u}\varphi^{(k-1)}(\boldsymbol{\theta}) + r_x(\tau)\left[P\varphi^{(k)}(t, u, x, 0) - \varphi^{(k)}(\boldsymbol{\theta})\right] + \frac{\partial}{\partial \tau}\varphi^{(k)}(\boldsymbol{\theta}).$$

Hence,

$$Q\varphi^{(k)}(\boldsymbol{\theta}) = \frac{\partial}{\partial t}\varphi^{(k-2)}(\boldsymbol{\theta}) - C(u, x)\frac{\partial}{\partial u}\varphi^{(k-1)}(\boldsymbol{\theta}). \quad [3.42]$$

For the specific $k = 2$ case, we have

$$Q\varphi^{(2)}(\boldsymbol{\theta}) = \frac{\partial}{\partial t}\varphi^{(0)}(\boldsymbol{\theta}) - C(u,x)\frac{\partial}{\partial u}\varphi^{(1)}(\boldsymbol{\theta})$$

$$= \frac{\partial}{\partial t}\varphi^{(0)}(\boldsymbol{\theta}) + C(u,x)\frac{\partial}{\partial u}\mathcal{R}_0 C(u,x)\frac{\partial}{\partial u}\varphi^{(0)}(\boldsymbol{\theta}) - C(u,x)\frac{\partial}{\partial u}n_1(\boldsymbol{\theta}).$$

Now, by applying the operator Π_1 on the left we obtain

$$\Pi_1 Q\varphi^{(2)}(\boldsymbol{\theta}) = 0 = \frac{\partial}{\partial t}\Pi_1\varphi^{(0)}(\boldsymbol{\theta}) - \Pi_1 C(u,x)\frac{\partial}{\partial u}\varphi^{(1)}(\boldsymbol{\theta}),$$

and taking into account that $\varphi^{(0)}(\boldsymbol{\theta}) \in ker(Q)$, i.e.

$$\Pi_1\varphi^{(0)}(\boldsymbol{\theta}) = \varphi^{(0)}(\boldsymbol{\theta}),$$

we obtain

$$\frac{\partial}{\partial t}\varphi^{(0)}(\boldsymbol{\theta}) + \Pi_1 C(u,x)\frac{\partial}{\partial u}\mathcal{R}_0 C(u,x)\frac{\partial}{\partial u}\Pi_1\varphi^{(0)}(\boldsymbol{\theta})$$

$$+\Pi_1 C(u,x)\frac{\partial}{\partial u}\Pi_1 n_1(\boldsymbol{\theta}) = 0.$$

Since the operator Π_1 does not depend on u, then

$$\Pi_1 C(u,x)\frac{\partial}{\partial u}\Pi_1 n_1(\boldsymbol{\theta})$$

is equivalent to

$$\Pi_1 C(u,x)\Pi_1\frac{\partial}{\partial u}n_1(\boldsymbol{\theta}) = 0,$$

and from the balance condition B5

$$\Pi_1 C(u,x)\Pi_1 = 0.$$

Therefore,

$$\frac{\partial}{\partial t}\varphi^{(0)}(\boldsymbol{\theta}) + \Pi_1 C(u,x)\frac{\partial}{\partial u}\mathcal{R}_0 C(u,x)\frac{\partial}{\partial u}\Pi_1\varphi^{(0)}(\boldsymbol{\theta}) = 0. \quad [3.43]$$

Similarly, it follows from [3.42] that

$$\varphi^{(k)}(\boldsymbol{\theta}) = \mathcal{R}_0\left[\frac{\partial}{\partial t}\varphi^{(k-2)}(\boldsymbol{\theta}) - C(u,x)\frac{\partial}{\partial u}\varphi^{(k-1)}(\boldsymbol{\theta})\right] + n_k(\boldsymbol{\theta}), \quad [3.44]$$

where $n_k(\boldsymbol{\theta}) \in ker(Q)$.

To find $n_k(\boldsymbol{\theta})$, we use the fact that $\varphi^{(0)}(\boldsymbol{\theta}) \in ker(Q)$, and we establish

$$n_0(\boldsymbol{\theta}) = \varphi^{(0)}(\boldsymbol{\theta}).$$

From [3.41] and [3.44], we have for $k = 2$,

$$\varphi^{(2)}(\boldsymbol{\theta}) = \mathcal{R}_0 \frac{\partial}{\partial t} n_0(\boldsymbol{\theta}) + \mathcal{R}_0 C(u,x) \frac{\partial}{\partial u} \mathcal{R}_0 C(u,x) \frac{\partial}{\partial u} n_0(\boldsymbol{\theta})$$

$$-\mathcal{R}_0 C(u,x) \frac{\partial}{\partial u} n_1(\boldsymbol{\theta}) + n_2(\boldsymbol{\theta}). \quad [3.45]$$

By letting $k = 3$ in [3.42], we obtain

$$Q\varphi^{(3)}(\boldsymbol{\theta}) = \frac{\partial}{\partial t}\varphi^{(1)}(\boldsymbol{\theta}) - C(u,x)\frac{\partial}{\partial u}\varphi^{(2)}(\boldsymbol{\theta}).$$

Then, by using [3.41] and [3.45], it follows that

$$Q\varphi^{(3)}(\boldsymbol{\theta}) = \frac{\partial}{\partial t}\varphi^{(1)}(\boldsymbol{\theta}) - C(u,x)\frac{\partial}{\partial u}\varphi^{(2)}(\boldsymbol{\theta})$$

$$= -\frac{\partial}{\partial t}\mathcal{R}_0 C(u,x)\frac{\partial}{\partial u} n_0(\boldsymbol{\theta}) + \frac{\partial}{\partial t} n_1(\boldsymbol{\theta}) - C(u,x)\frac{\partial}{\partial u}\mathcal{R}_0\frac{\partial}{\partial t} n_0(\boldsymbol{\theta})$$

$$-C(u,x)\frac{\partial}{\partial u}\mathcal{R}_0 C(u,x)\frac{\partial}{\partial u}\mathcal{R}_0 C(u,x)\frac{\partial}{\partial u} n_0(\boldsymbol{\theta})$$

$$+C(u,x)\frac{\partial}{\partial u}\mathcal{R}_0 C(u,x)\frac{\partial}{\partial u} n_1(\boldsymbol{\theta}) - C(u,x)\frac{\partial}{\partial u} n_2(\boldsymbol{\theta}). \quad [3.46]$$

Taking into account the balance condition C5, and multiplying [3.46] by Π_1, we have

$$\left(\frac{\partial}{\partial t} + \Pi_1 C(u,x)\frac{\partial}{\partial u}\mathcal{R}_0 C(u,x)\Pi_1\frac{\partial}{\partial u}\right) n_1(\boldsymbol{\theta})$$

$$-\left(\Pi_1 C(u,x)\frac{\partial}{\partial u}\mathcal{R}_0\frac{\partial}{\partial t}\Pi_1 + \Pi_1\frac{\partial}{\partial t}\mathcal{R}_0 C(u,x)\Pi_1\frac{\partial}{\partial u}\right.$$

$$\left. + \Pi_1 C(u,x)\frac{\partial}{\partial u}\mathcal{R}_0 C(u,x)\frac{\partial}{\partial u}\mathcal{R}_0 C(u,x)\Pi_1\frac{\partial}{\partial u}\right) n_0(\boldsymbol{\theta}) = 0. \quad [3.47]$$

After solving [3.47], we can express $n_1(\boldsymbol{\theta})$ in terms of $n_0(\boldsymbol{\theta})$.

Now, we consider [3.42] for $k = 4$, namely,

$$Q\varphi^{(4)}(\boldsymbol{\theta}) = \frac{\partial}{\partial t}\varphi^{(2)}(\boldsymbol{\theta}) - C(u,x)\frac{\partial}{\partial u}\varphi^{(3)}(\boldsymbol{\theta}). \quad [3.48]$$

Let us substitute

$$\varphi^{(3)}(\boldsymbol{\theta}) = \mathcal{R}_0 \left[\frac{\partial}{\partial t} \varphi^{(1)}(\boldsymbol{\theta}) - C(u,x) \frac{\partial}{\partial u} \varphi^{(2)}(\boldsymbol{\theta}) \right] + n_3(\boldsymbol{\theta})$$

and $\varphi^{(2)}(\boldsymbol{\theta})$ from [3.45] into [3.48], and then multiply by Π_1 both sides of the resulting equation.

In the same manner, we use the balance condition B5. Then we obtain a differential equation that relates $n_1(\boldsymbol{\theta})$ and $n_2(\boldsymbol{\theta})$. After solving this equation, we can express $n_2(\boldsymbol{\theta})$ in terms of $n_1(\boldsymbol{\theta})$. The same procedure can be applied to obtain $n_i(\boldsymbol{\theta})$ for all $i = 0, 1, 2, \ldots$ as follows:

$$\begin{aligned}
&\left(\frac{\partial}{\partial t} + \Pi_1 C(u,x) \frac{\partial}{\partial u} \mathcal{R}_0 C(u,x) \Pi_1 \frac{\partial}{\partial u} \right) n_{i+1}(\boldsymbol{\theta}) \\
&- \Bigg(\Pi_1 C(u,x) \frac{\partial}{\partial u} \mathcal{R}_0 \frac{\partial}{\partial t} \Pi_1 + \Pi_1 \frac{\partial}{\partial t} \mathcal{R}_0 C(u,x) \Pi_1 \frac{\partial}{\partial u} \\
&+ \Pi_1 C(u,x) \frac{\partial}{\partial u} \mathcal{R}_0 C(u,x) \frac{\partial}{\partial u} \mathcal{R}_0 C(u,x) \Pi_1 \frac{\partial}{\partial u} \Bigg) n_i(\boldsymbol{\theta}) = 0.
\end{aligned}$$

By solving [3.43] and substituting $\varphi^{(0)}(\boldsymbol{\theta})$, $\varphi^{(1)}(\boldsymbol{\theta})$ into [3.42], we obtain $\varphi^{(2)}(\boldsymbol{\theta})$.

Continuing the iterative process in equation [3.42], we obtain $\varphi^{(k)}(\boldsymbol{\theta})$ for all $k = 0, 1, 2, \ldots$.

Then, regarding the first singular term, we have

$$\frac{\partial}{\partial t} g^{(1)} = Q g^{(1)}, \qquad [3.49]$$

and for $k \geq 1$, we have

$$\frac{\partial}{\partial t} g^{(k+1)} - Q g^{(k+1)} = C(u,x) \frac{\partial}{\partial u} g^{(k)}. \qquad [3.50]$$

By solving [3.44], we obtain

$$g^{(1)}(\boldsymbol{\theta}) = \exp_0 \{Qt\} \, g^{(1)}(0, u, x, \tau),$$

where the modified exponent $\exp_0 \{Qt\} = \exp \{Qt\} - \Pi_1$ satisfies

$$\lim_{t \to \infty} \|\exp \{Qt\} - \Pi_1\| = 0.$$

Thus, $\lim_{t\to\infty}\left\|g^{(1)}(\boldsymbol{\theta})\right\|=0$.

It is easily verified that the semigroup of operators of the Markov process $(\xi(t),\tau(t))$ is a solution of equation [3.49].

By solving equation [3.50], we have

$$g^{(k+1)}(\boldsymbol{\theta})=\exp_0\{Qt\}\,g^{(k+1)}(0,u,x,\tau)$$

$$+\int_0^t\exp_0\{Q(t-s)\}C(u,x)\frac{\partial}{\partial u}g^{(k)}(s,u,x,\tau)\,ds.$$

Thus, $g^{(k)}$ can be obtained from [3.50] recursively for all $k\geq 1$, and the theorem is proved.

3.5. Asymptotic expansion for distribution of random motion in Markov media under the Kac condition

In this section, we study an asymptotic expansion for the distribution of the random motion of a particle driven by a Markov process under a diffusion approximation. We show that the singularly perturbed equation of a Markov random motion can be reduced to the regularly perturbed equation for the distribution of the random motion.

3.5.1. *The equation for the probability density of the particle position performing a random walk in* $\mathbb{R}^n$

Let us consider the random motion of a particle in $\mathbb{R}^n$ driven by a Markov process $\{\xi(t)\}$, whose sojourn times at states are exponentially distributed with rate $\lambda>0$ and transition probabilities $p_{ij}=\frac{1}{2n-1}(1-\delta_{ij})$, $i,j\in E=\{1,2,\ldots,2n\}$, where E is the phase space of $\xi(t)$.

Let $\mathbf{e}_1,\ldots,\mathbf{e}_n$ be a Cartesian basis of $\mathbb{R}^n$. Define $\mathbf{e}_{n+k}=-\mathbf{e}_k$, $k=1,2,\ldots,n$ and $\mathbf{v}_i=v\,\mathbf{e}_i$, $i=1,2,\ldots,2n$, where $v>0$ is the constant speed of the particle.

We assume that the particle moves in an n-dimensional space in the following manner: If at some instant t the particle has velocity $\mathbf{v}_i$, then at a renewal epoch of the Markov process the particle takes a new velocity $\mathbf{v}_j$, $j\neq i$ with probability $p_{ij}=\frac{1}{2n-1}$. The particle continues its motion with velocity $\mathbf{v}_j$ until the next renewal occurrence of the Markov process, and so on.

Let us denote by $\mathbf{r}(t)=(x_1(t),\,x_2(t),\ldots,\,x_n(t))$, $t\geq 0$ the particle position at time t.

Consider the function

$$\mathbf{C}(i) = (C_1(i), C_2(i), \ldots, C_n(i)) = \mathbf{v}_i, \; i \in E.$$

Then, the position of the particle at time t can be expressed as

$$\mathbf{r}(t) = \mathbf{r}(0) + \int_0^t \mathbf{C}(\xi(t))\, dt.$$

3.5.2. *Equation for the probability density of the particle position*

Let us consider a bivariate stochastic process $\varsigma(t) = (\mathbf{r}(t), \xi(t))$ with the phase space $\mathbb{R}^n \times E$. It is well-known that this process is a Markov process and that it has the generating operator (Korolyuk 1993; Korolyuk and Korolyuk 1999):

$$A\varphi(\mathbf{r}, i) = \mathbf{C}(i)\varphi'(\mathbf{r}, i) + \lambda [P\varphi(\mathbf{r}, i) - \varphi(\mathbf{r}, i)], \quad i \in E, \qquad [3.51]$$

where

$$\mathbf{C}(i)\varphi'(\mathbf{r}, i) = C_1(i)\frac{\partial}{\partial x}\varphi(\mathbf{r}, i) + C_2(i)\frac{\partial}{\partial y}\varphi(\mathbf{r}, i) + \ldots + C_n(i)\frac{\partial}{\partial z}\varphi(\mathbf{r}, i)$$

and

$$P\varphi(\mathbf{r}, i) = \frac{\lambda}{2n-1} \sum_{\substack{j \in E \\ j \neq i}} \varphi(\mathbf{r}, j).$$

Now, let us consider the density function

$$\begin{aligned} &f_i(t, x_1, \ldots, x_n)\, dx_1 \ldots dx_n \\ &= P\{x_1 \le x_1(t) \le x_1 + dx_1, \ldots, x_n \le x_n(t) \le x_n + dx_n;\; \xi(t) = i\}. \end{aligned}$$

It is easily verified that

$$f(t, x_1, \ldots, x_n) = \sum_{i=1}^{n} f_i(t, x_1, \ldots, x_n),$$

is the probability density of the particle position in $\mathbb{R}^n$ at time t.

LEMMA 3.4.– The function f satisfies the following differential equation:

$$\prod_{k=1}^{2n} \left\{ \frac{\partial}{\partial t} + (-1)^k v \frac{\partial}{\partial x_k} + \frac{2n\lambda}{2n-1} \right\} f$$

$$+\frac{2n\lambda}{2n-1}\sum_{l=1}^{2n}\prod_{\substack{k=1\\k\neq l}}^{2n}\left\{\frac{\partial}{\partial t}+(-1)^k v\frac{\partial}{\partial x_k}+\frac{2n\lambda}{2n-1}\right\}f=0.$$

PROOF.– For $i\in E$, the function f_i satisfies the first Kolmogorov equation, namely

$$\frac{\partial}{\partial t}f_i\left(t,x_1,\ldots,x_n\right)=Af_i\left(t,x_1,\ldots,x_n\right),\quad i\in E, \qquad [3.52]$$

with initial conditions $f_i\left(0,\, x_1,\ldots,\, x_n\right)=f_i^{(0)}$.

Equation [3.52] can be written in more detail as follows:

$$\frac{\partial f_i\left(t,\, x_1,\ldots,\, x_n\right)}{\partial t}+(-1)^i\frac{\partial f_i\left(t,\, x_1,\ldots,x_n\right)}{\partial x_i}+\lambda f_i\left(t,\, x_1,\ldots,\, x_n\right)$$

$$-\frac{\lambda}{2n-1}\sum_{j\in E\setminus i}f_j\left(t,x_1,\ldots,x_n\right)=0,\quad i\in E. \qquad [3.53]$$

Let us define

$$\mathbf{f}\left(t,\, x_1,\ldots,\, x_n\right)=\left\{f_i\left(t,x_1,\ldots,\, x_n\right),\quad i\in E\right\}.$$

The set of equations [3.53] can be written in the following form:

$$L_{2n}\mathbf{f}=0,$$

where for $k\neq i,\ k,i\in E$

$$L_{2n}=\left\|l_{ij}\right\|_{i,j\in E},\ l_{kk}=\frac{\partial}{\partial t}+(-1)^k v\frac{\partial}{\partial x_k}+\lambda,\ l_{ik}=\frac{-\lambda}{2n-1}.$$

The function f satisfies the following equation:

$$\det\left(L_{2n}\right)f=0. \qquad [3.54]$$

with the initial condition $f\left(0,\, x_1,\ldots,\, x_n\right)=f_0=\sum_{i=1}^n f_i^{(0)}$.

The determinant of matrix L_{2n} is well-known and has the form

$$\det\left(L_{2n}\right)=\prod_{k=1}^{2n}\left(\frac{\partial}{\partial t}+(-1)^k v\frac{\partial}{\partial x_k}+\frac{2n\lambda}{2n-1}\right)$$

$$+\frac{2n\lambda}{2n-1}\sum_{l=1}^{2n}\prod_{\substack{k=1\\k\neq l}}^{2n}\left(\frac{\partial}{\partial t}+(-1)^k v\frac{\partial}{\partial x_k}+\frac{2n\lambda}{2n-1}\right).$$

Since $v\frac{\partial}{\partial x_k}$ and $-v\frac{\partial}{\partial x_k}$ appear symmetrically in L_{2n}, it is easy to see that all monomials of the polynomial $\det\left(L_{2n}\right)$ contain v^k only for even powers of $k\geq 0$.

3.5.3. *Reduction of a singularly perturbed evolution equation to a regularly perturbed equation*

Let us define $\lambda = \varepsilon^{-2}$, $v = c\varepsilon^{-1}$, where $\varepsilon > 0$ is a small parameter. It is well-known (Kac 1974) that the solution of equation [3.54] in the hydrodynamical limit (as $\varepsilon \to 0$) weakly converges to the corresponding functional of Wiener process.

As mentioned before, we can find an asymptotic expansion of the solution of equation [3.54], and it consists of regular and singular terms. However, this technique may involve tedious calculations. So, we will follow an alternative approach here.

PROPOSITION 3.1.– The equation $\det (L_{2n}) f = 0$ is regular perturbed, i.e. if we multiply it by ε^{4n-2}, then we obtain

$$\frac{\partial}{\partial t} f = \frac{2n-1}{2n^2}\left(\frac{\partial^2}{\partial x_1^2} + \ldots + \frac{\partial^2}{\partial x_n^2}\right) f + D_\varepsilon f, \qquad [3.55]$$

where $D_\varepsilon = \varepsilon^2 D_1 + \varepsilon^4 D_2 + \ldots$, and D_i, $i = 1, 2, \ldots$, are respective differential operators.

PROOF.– To avoid cumbersome expressions, we consider the case when $n = 3$. Let us define $x = x_1$, $y = x_2$, $z = x_3$, and the coordinates x, y, z as ψ in this section. In this case, equation [3.53] has the following form:

$$\frac{\partial}{\partial t} f_i(t,\psi) + (-1)^i v \frac{\partial}{\partial x} f_i(t,\psi) + \lambda f_i(t,\psi)$$
$$-\frac{\lambda}{5} \sum_{\substack{j \in E \\ j \neq i}} f_j(t,\psi) = 0, \quad i = 1, 2, \ldots, n. \qquad [3.56]$$

Now $v = \varepsilon^{-1}$ and $\lambda = \varepsilon^{-2}$, we obtain the following singularly perturbed system of equations

$$\frac{\partial}{\partial t} f_i(t,\psi) - \varepsilon^{-1} \frac{\partial}{\partial x} f_i(t,\psi) + \varepsilon^{-2} f_i(t,\psi)$$
$$-\varepsilon^{-2}\frac{1}{5} \sum_{\substack{j \in E \\ j \neq i}} f_j(t,\psi) = 0, \qquad i = 1, 2, \ldots, n.$$

Let us consider $\det (L_6) f = 0$, where

$$f(t, x_1, \ldots, x_n) = \sum_{i=1}^{n} f_i(t, x_1, \ldots, x_n).$$

It is easy to see that elements of matrix $L_6 = (l_{ij})_{i,j\in E}$ are as follows: $l_{ii} = \frac{\partial}{\partial t} + (-1)^i v \frac{\partial}{\partial x_i} + \lambda$ and $l_{ij} = \frac{-\lambda}{5}$, $i \neq j, i, j \in \{1, 2, \cdots, 6\}$.

Hence, the equation $\det(L_6) f = 0$ becomes of the following form:

$$\det(L_6) f(t,\psi) = \left\{ \frac{7776}{3125}\varepsilon^{-10}\frac{\partial}{\partial t} - \frac{432}{625}\varepsilon^{-10}\Delta - \frac{432}{125}\varepsilon^{-8}\Delta\frac{\partial}{\partial t} \right.$$

$$- \frac{144}{25}\varepsilon^{-6}\Delta\frac{\partial^2}{\partial t^2} + \frac{24}{25}\varepsilon^{-8}\Delta^{(2)} + \frac{72}{5}\varepsilon^{-4}\frac{\partial^4}{\partial t^4} - 4\varepsilon^{-4}\Delta\frac{\partial^3}{\partial t^3} + 2\varepsilon^{-6}\Delta^{(2)}\frac{\partial}{\partial t}$$

$$+6\varepsilon^{-2}\frac{\partial^5}{\partial t^5} + \frac{\partial^6}{\partial t^6} + \varepsilon^{-8}\Delta^{(2)}\frac{\partial^2}{\partial t^2} - \varepsilon^{-2}\Delta\frac{\partial^4}{\partial t^4} - \varepsilon^{-6}\frac{\partial^6}{\partial x^2\partial y^2\partial z^2}$$

$$\left. + \frac{1296}{125}\varepsilon^{-2}\frac{\partial^2}{\partial t^2} + \frac{432}{25}\varepsilon^{-6}\frac{\partial^3}{\partial t^3} \right\} f(t,\psi) = 0, \qquad [3.57]$$

with the initial condition $f(0,\psi) = f_0$, where

$$\Delta = \frac{\partial^2}{\partial x^2} + \frac{\partial^2}{\partial y^2} + \frac{\partial^2}{\partial z^2}, \qquad \Delta^{(2)} = \frac{\partial^4}{\partial x^2\partial y^2} + \frac{\partial^4}{\partial y^2\partial z^2} + \frac{\partial^4}{\partial x^2\partial z^2}.$$

Multiplying [3.57] by ε^{10}, we obtain

$$\left\{ \frac{\partial}{\partial t} - \frac{5}{18}\Delta + \varepsilon^2\frac{25}{6}\left[\frac{\partial^2}{\partial t^2} - \frac{1}{3}\frac{\partial}{\partial t}\Delta + \frac{5}{54}\Delta^{(2)}\right] \right.$$

$$+\varepsilon^4\frac{125}{18}\left[\frac{\partial^3}{\partial t^3} - \frac{1}{3}\frac{\partial^2}{\partial t^2}\Delta + \frac{5}{216}\frac{\partial}{\partial t}\Delta^{(2)}\right]$$

$$+\varepsilon^6\frac{625}{216}\left[\frac{\partial^4}{\partial t^4} - \frac{5}{18}\frac{\partial^3}{\partial t^3}\Delta + \frac{5}{72}\frac{\partial^2}{\partial t^2}\Delta^{(2)}\right]$$

$$\left. + \varepsilon^8\frac{3125}{1296}\left[\frac{\partial^5}{\partial t^5} - \frac{1}{6}\frac{\partial^5}{\partial t^4}\Delta^{(2)}\right] + \varepsilon^{10}\frac{3125}{7776}\frac{\partial^6}{\partial t^6} \right\} f = 0. \qquad [3.58]$$

LEMMA 3.5.– The solution of equation [3.58] with the initial condition

$$f(0,\psi) = u_0(0,\psi) + \varepsilon^2 u_1(0,\psi) + \varepsilon^4 u_2(0,\psi) + \ldots$$

has the following asymptotic solution:

$$f(t,\psi) = u_0(t,\psi) + \varepsilon^2 u_1(t,\psi) + \varepsilon^4 u_2(t,\psi) + \ldots \qquad [3.59]$$

where the main term $u_0(t,\psi)$ represents the solution of equation

$$\frac{\partial}{\partial t}u_0(t,\psi) = \frac{5}{18}\Delta u_0(t,\psi).$$

PROOF.– To find the asymptotic expansion of the solution of equation [3.58], we apply the method in the proof of theorem 3.6. In conformity with this method, the solution of [3.58] can be expanded into the series [3.59], where $\varepsilon > 0$ is a small parameter.

Substituting [3.59] into [3.59], we can obtain the following equations for the calculation of u_i, $i \geq 0$:

$$\frac{\partial}{\partial t} u_0 (t, \psi) = \frac{5}{18} \Delta u_0(t, \psi);$$

$$\frac{\partial}{\partial t} u_1 (t, \psi) = \frac{5}{18} \Delta u_1 (t, \psi) + \frac{25}{6} \left[\frac{\partial^2}{\partial t^2} - \frac{1}{3} \frac{\partial}{\partial t} \Delta + \frac{5}{54} \Delta^{(2)} \right] u_0(t, \psi);$$

$$\frac{\partial}{\partial t} u_2 (t, \psi) = \frac{5}{18} \Delta u_2 (t, \psi) + \frac{25}{6} \left[\frac{\partial^2}{\partial t^2} - \frac{1}{3} \frac{\partial}{\partial t} \Delta + \frac{5}{54} \Delta^{(2)} \right] u_1 (t, \psi)$$

$$+ \frac{125}{18} \left[\frac{\partial^3}{\partial t^3} - \frac{1}{3} \frac{\partial^2}{\partial t^2} \Delta + \frac{5}{216} \frac{\partial}{\partial t} \Delta^{(2)} \right] u_0(t, \psi);$$

$$\vdots$$

$$\frac{\partial}{\partial t} u_{m+5} (t, \psi) = \frac{5}{18} \Delta u_{m+5} (t, \psi)$$

$$+ \frac{25}{6} \left[\frac{\partial^2}{\partial t^2} - \frac{1}{3} \frac{\partial}{\partial t} \Delta + \frac{5}{54} \Delta^{(2)} \right] u_{m+4} (t, \psi)$$

$$+ \frac{125}{18} \left[\frac{\partial^3}{\partial t^3} - \frac{1}{3} \frac{\partial^2}{\partial t^2} \Delta + \frac{5}{216} \frac{\partial}{\partial t} \Delta^{(2)} \right] u_{m+3} (t, \psi)$$

$$+ \frac{625}{216} \left[\frac{\partial^4}{\partial t^4} - \frac{5}{18} \frac{\partial^3}{\partial t^3} \Delta + \frac{5}{72} \frac{\partial^2}{\partial t^2} \Delta^{(2)} \right] u_{m+2} (t, \psi)$$

$$+ \frac{3125}{1296} \left[\frac{\partial^5}{\partial t^5} - \frac{1}{6} \frac{\partial^5}{\partial t^4} \Delta^{(2)} \right] u_{m+1} (t, \psi) + \frac{3125}{7776} \frac{\partial^6}{\partial t^6} u_m (t, \psi) = 0,$$

for $m \geq 0$.

Let us consider

$$f_k^{\varepsilon} (t, \psi) = u_0 (t, \psi) + \varepsilon^2 u_1 (t, \psi) + \ldots + \varepsilon^{2k} u_k(t, \psi).$$

The solution of a singularly perturbed equation of type [3.53] and the remainder of the asymptotic expansion, in the context of diffusion approximation, was studied in Korolyuk and Turbin (1993). Taking into account that

$$f (t, \psi) = \sum_{i \in E} f_i (t, \psi),$$

it follows from the estimate of the reminder given in Korolyuk and Turbin (1993) that

$$\|f(t,\psi) - f_n^{\varepsilon}(t,\psi)\| = O\left(\varepsilon^k\right).$$

3.6. Asymptotic estimation for application of the telegraph process as an alternative to the diffusion process in the Black–Scholes formula

In this section, we study the one-dimensional transport process in the case of disbalance. In the hydrodynamic limit, this process approximates the diffusion process on the line. The application of the telegraph process for option prices was studied in Ratanov (2007). By using the asymptotic expansion for the singularly perturbed random evolution in Markov media with no balance condition, we study the accuracy of the application of the telegraph process in the Black–Scholes formula, as an alternative to the diffusion process.

3.6.1. *Asymptotic expansion for the singularly perturbed random evolution in Markov media in case of disbalance*

Let $\{\xi(t),\, t \geq 0\}$ be a Markov process on the phase space $\{0,1\}$ with the infinitesimal matrix

$$Q = \lambda \begin{pmatrix} -1 & 1 \\ 1 & -1 \end{pmatrix}.$$

Now, consider the random velocity model or transport process

$$x(t) = x_0 + \int_0^t v(s)ds,$$

where

$$v(t) = \begin{cases} v_0, & \text{when } \xi(t) = 0; \\ v_1, & \text{when } \xi(t) = 1. \end{cases}$$

The infinitesimal operator A of the bivariate process

$$\{\varsigma(t) = (x(t), \xi(t)),\, t \geq 0\}$$

is of the following form:

$$A\varphi(x,0) = v_0 \frac{\partial}{\partial x}\varphi(x,0) + \lambda\varphi(x,1) - \lambda\varphi(x,0)$$

$$A\varphi(x,1) = v_1 \frac{\partial}{\partial x}\varphi(x,1) + \lambda\varphi(x,0) - \lambda\varphi(x,1), \qquad [3.60]$$

where $\varphi \in D(A)$ is the domain of the operator A and $x \in \mathbb{R}$.

We can interpret this operator in the following equivalent manner: assume $\mathbb{Z} = \mathbb{R} \times \{0, 1\}$ and

$$T_t\varphi(x,i) = \int_{\mathbb{R}} \varphi(z) P\{\varsigma(t) \in dz \mid \varsigma(0) = (x,i)\}, \quad i \in \{0,1\}.$$

Then,

$$A\varphi(x,i) = \lim_{\triangle t \to 0+} \frac{T_{\triangle t}\varphi(x,i) - \varphi(x,i)}{\triangle t}.$$

Let us define $u_i(x,t) = T_t\varphi(x,i)$. Thus,

$$\frac{\partial}{\partial t} u_i(x,t) = AT_t\varphi(x,i) = Au_i(x,t).$$

Now, let us consider the scaled evolution equation

$$x_\varepsilon(t) = x_0 + \frac{1}{\varepsilon}\int_0^t v\left(\frac{s}{\varepsilon^2}\right) ds,$$

with corresponding velocities v_i/ε , $i \in \{0,1\}$. Since the Markov process $v\left(\frac{t}{\varepsilon^2}\right)$ has scaled time, then its infinitesimal operator is of the form $\frac{1}{\varepsilon^2}Q$.

Hence, we obtain the following system of Kolmogorov backward differential equations

$$\frac{\partial}{\partial t}u_0^\varepsilon(x,t) = \frac{v_0}{\varepsilon}\frac{\partial}{\partial x}u_0^\varepsilon(x,t) + \frac{\lambda}{\varepsilon^2}u_1^\varepsilon(x,t) - \frac{\lambda}{\varepsilon^2}u_0^\varepsilon(x,t)$$

$$\frac{\partial}{\partial t}u_1^\varepsilon(x,t) = \frac{v_1}{\varepsilon}\frac{\partial}{\partial x}u_1^\varepsilon(x,t) + \frac{\lambda}{\varepsilon^2}u_0^\varepsilon(x,t) - \frac{\lambda}{\varepsilon^2}u_1^\varepsilon(x,t). \quad [3.61]$$

Equations [3.61] can be written in matrix form as follows:

$$\frac{\partial}{\partial t}\mathbf{u}^\varepsilon(x,t) = \frac{1}{\varepsilon}V\nabla\mathbf{u}^\varepsilon(x,t) + \frac{1}{\varepsilon^2}Q\mathbf{u}^\varepsilon(x,t) \quad [3.62]$$

where

$$\mathbf{u}^\varepsilon(x,t) = \begin{pmatrix} u_0^\varepsilon(x,t) \\ u_1^\varepsilon(x,t) \end{pmatrix}, \qquad V = \begin{pmatrix} v_0 & 0 \\ 0 & v_1 \end{pmatrix},$$

and

$$V\nabla = \begin{pmatrix} v_0\frac{\partial}{\partial x} & 0 \\ 0 & v_1\frac{\partial}{\partial x} \end{pmatrix}, \qquad Q = \lambda\begin{pmatrix} -1 & 1 \\ 1 & -1 \end{pmatrix}.$$

Let us define

$$R_0 = \int_0^{\infty} \left(\Pi - e^{Qt}\right) dt,$$

where $e^{Qt} = \{p_{ij}(t);\ i,j \in \{0,1\}\}$ is the set of time-dependent transition probabilities, and

$$\Pi = \begin{pmatrix} \frac{1}{2} & \frac{1}{2} \\ \frac{1}{2} & \frac{1}{2} \end{pmatrix}.$$

is the projector operator on $N(Q) = ker(Q)$ (Volume 1, Chapter 1, section 1.2).

Thus, $R_0 = \Pi - (Q + \Pi)^{-1}$ is the potential operator of $\xi(t)$ (Volume 1, Chapter 1, section 1.3) (Korolyuk and Turbin 1976). The balance condition states that $\frac{v_0+v_1}{2} = 0$, i.e. $v_1 = -v_0$. This balance condition can be also expressed as $\Pi V \Pi = 0$ (see condition B5 of theorem 3.6).

Now, we will consider the following disbalance condition: $v_0 = v + \Delta_1$ and $v_1 = -v - \Delta_2$, where $\Delta_i = \varepsilon a_i$, $i = 1,2$, $\varepsilon > 0$. It is easily verified that the infinitesimal operator of the process $\zeta_\varepsilon(t) = \left(x_\varepsilon(t),\ \xi\left(\frac{t}{\varepsilon^2}\right)\right)$ is of the following form:

$$A_\varepsilon = \frac{1}{\varepsilon} V\nabla + A\nabla + \frac{1}{\varepsilon^2} Q,$$

where

$$A = \begin{pmatrix} a_1 & 0 \\ 0 & a_2 \end{pmatrix}.$$

Denote as $u_i^\varepsilon(x,t) = T_t^\varepsilon \varphi(x,i) = \int_Z \varphi(z) P\{\zeta_\varepsilon(t) \in dz / \zeta_\varepsilon(0) = (x,i)\}$.

Then, similarly as above, we can write the matrix equation

$$\frac{\partial}{\partial t}\mathbf{u}^\varepsilon(x,t) = \frac{1}{\varepsilon} V\nabla \mathbf{u}^\varepsilon(x,t) + A\nabla \mathbf{u}^\varepsilon(x,t) + \frac{1}{\varepsilon^2} Q\mathbf{u}^\varepsilon(x,t). \quad [3.63]$$

By using the technique of asymptotic expansion, as in the proof of theorem 3.6, we will find a solution of equation [3.63] in the following form:

$$\mathbf{u}^\varepsilon(x,t) = \mathbf{u}^{(0)}(x,t) + \sum_{n=1}^{\infty} \varepsilon^n \left(\mathbf{u}^{(n)}(x,t) + \mathbf{v}^{(n)}(x, t/\varepsilon^2)\right), \quad [3.64]$$

where $\mathbf{u}^{(n)}(x,t)$, $n = 0,1,2,\ldots$ are the regular terms of the expansion, whereas $\mathbf{v}^{(n)}(x, t/\varepsilon^2)$, $n = 1,2,\ldots$ are the singular ones.

Then, by substituting equation [3.64] into [3.63], we obtain

$$Q\mathbf{u}^{(0)}(x,t) = 0, \qquad Q\mathbf{u}^{(1)}(x,t) + V\nabla\mathbf{u}^{(0)}(x,t) = 0,$$

$$Q\mathbf{u}^{(2)}(x,t) + V\nabla\mathbf{u}^{(1)}(x,t) + A\nabla\mathbf{u}^{(0)}(x,t) - \frac{\partial}{\partial t}\mathbf{u}^{(0)}(x,t) = 0, \qquad [3.65]$$

$$Q\mathbf{u}^{(k+2)}(x,t) + V\nabla\mathbf{u}^{(k+1)}(x,t) + A\nabla\mathbf{u}^{(k)}(x,t) - \frac{\partial}{\partial t}\mathbf{u}^{(k)}(x,t) = 0,$$

for $k \geq 0$.

Hence, $\mathbf{u}^{(0)}(x,t) \in N(Q)$, i.e. $\Pi\mathbf{u}^{(0)}(x,t) = \mathbf{u}^{(0)}(x,t)$.

It follows from equation [3.65] that

$$\mathbf{u}^{(1)}(x,t) = R_0 V\nabla\mathbf{u}^{(0)}(x,t) + c_1(t), \qquad [3.66]$$

where $c_1(t) \in N(Q)$.

Similarly,

$$Q\mathbf{u}^{(2)}(x,t) = \frac{\partial}{\partial t}\mathbf{u}^{(0)}(x,t) - V\nabla\mathbf{u}^{(1)}(x,t)$$

$$= \frac{\partial}{\partial t}\mathbf{u}^{(0)}(x,t) - V\nabla R_0 V\nabla\mathbf{u}^{(0)}(x,t) - V\nabla c_1(t) - A\nabla\mathbf{u}^{(0)}(x,t). \qquad [3.67]$$

Multiplying equation [3.67] by the operator Π, we obtain

$$\Pi Q\mathbf{u}^{(2)}(x,t) = 0$$

$$= \frac{\partial}{\partial t}\mathbf{u}^{(0)}(x,t) - \Pi V\nabla R_0 V\nabla\Pi\mathbf{u}^{(0)}(x,t) - \Pi A\nabla\Pi\mathbf{u}^{(0)}(x,t). \qquad [3.68]$$

Here we use the fact that $\Pi V\nabla c_1(t) = \Pi V\nabla\Pi c_1(t) = 0$.

Therefore, the main term of the expansion [3.64] satisfies the diffusion equation [3.68].

We can obtain similar equations for the rest of the terms, but we are just interested in the main term $\mathbf{u}^{(0)}(x,t)$.

Now, let us write the matrix equation [3.63] in the following form:

$$\begin{pmatrix} \frac{\partial}{\partial t} - \frac{v}{\varepsilon}\frac{\partial}{\partial x} - a_1\frac{\partial}{\partial x} + \frac{\lambda}{\varepsilon^2} & \frac{\lambda}{\varepsilon^2} \\ \frac{\lambda}{\varepsilon^2} & \frac{\partial}{\partial t} + \frac{v}{\varepsilon}\frac{\partial}{\partial x} + a_2\frac{\partial}{\partial x} + \frac{\lambda}{\varepsilon^2} \end{pmatrix} \mathbf{u}^{(\varepsilon)}(x,t)$$

$$= \Psi\mathbf{u}^{(\varepsilon)}(x,t) = 0.$$

It is easy to verify that the function

$$f_\varepsilon(x,t) = u_0^\varepsilon(x,t) + u_1^\varepsilon(x,t)$$

is the solution of the following equation

$$\det(\Psi)\, f_\varepsilon(x,t) = 0. \qquad [3.69]$$

Writing equation [3.69] in more detail, we have

$$\left[\left(2\lambda\frac{\partial}{\partial t} - v^2\frac{\partial^2}{\partial x^2} + \lambda\left(a_2 - a_1\right)\frac{\partial}{\partial x}\right)\right.$$
$$\left. + \varepsilon^2\left(\frac{\partial^2}{\partial t^2} + \left(a_2 - a_1\right)\frac{\partial^2}{\partial x \partial t} - \frac{v\left(a_2 + a_1\right)}{\varepsilon}\frac{\partial^2}{\partial x^2} - a_1 a_2 \frac{\partial^2}{\partial x^2}\right)\right]$$
$$\times f_\varepsilon(x,t) = 0. \qquad [3.70]$$

Let us define the notation $u^{(0)}(x,t) = (u_0(x,t),\, u_1(x,t))$and $f_0(x,t) = u_0(x,t) + u_1(x,t)$. Since $\mathbf{u}^\varepsilon\,(x,t) > \mathbf{u}^{(0)}(x,t)$ as $\varepsilon > 0$ (Korolyuk and Korolyuk 1999), we have $\lim_{\varepsilon\to 0} f_\varepsilon(x,t) = f_0(x,t)$.

Hence, it follows from equation [3.70] that

$$\left(\frac{\partial}{\partial t} - \frac{v^2}{2\lambda}\frac{\partial^2}{\partial x^2} + \frac{a_2 - a_1}{2}\frac{\partial}{\partial x}\right) f_0(x,t) = 0. \qquad [3.71]$$

So, if $a_2 \neq a_1$, then $f_0(x,t)$ satisfies the diffusion equation [3.71] with drift coefficient $\frac{a_1 - a_2}{2}$ and diffusion coefficient $\frac{v^2}{2\lambda}$.

3.6.2. *Application to an economic model of stock market*

The well-known Black–Scholes formula (Dineen 2005) gives the price X_t of a stock at time t such that

$$X_t = X_0 \exp(\mu t + \sigma W_t), \qquad [3.72]$$

where μ is the drift, σ is the volatility of the stock, and $\{W_t\}$ is the Wiener process.

By substituting $\mu = \frac{a_1 - a_2}{2}$ and $\sigma = \frac{v}{\sqrt{2\lambda}}$ in equation [3.69], we obtain the diffusion equation for the process $\{\mu t + \sigma W_t\}$. Therefore, the Black–Scholes formula is obtained by considering an exponential Brownian motion for the share price X_t.

The Wiener process provides a mathematical consistent model for Brownian motion. However, it has the following drawbacks in capturing the physics of many applications:

1) modulus of the velocity is almost always infinite at any instant in time;

2) free path length of zero;

3) the path function of a particle is almost surely non-differentiable at any given point, and its Hausdorff dimension is equal to 1.5, i.e. the path function is fractal.

However, the actual movement of a physical particle as well as the actual evolution of share prices X_t are barely justified as fractal quantities. Taking into account these considerations, we propose to use the process given by equation [3.70] instead of the diffusion process given in equation [3.71] for most applications. This recommendation is based on the fact that this process does not suffer from the drawbacks mentioned above. For instance, the process in equation [3.70] has almost always a finite modulus of velocity $V_\varepsilon = \frac{v}{\varepsilon}$, non-zero free paths that depend on $\Lambda_\varepsilon = \frac{\lambda}{\varepsilon^2}$, and its trajectories are almost surely differentiable.

Some difficulties may arise in calculating the coefficients V_ε and Λ_ε. For such a purpose, we suggest considering the time between two consecutive renewal epochs (or, equivalently, two consequent impacts of a particle or two consequent changes of price) as an exponential distributed random variable with parameter Λ_ε. Then, Λ_ε can be estimated from experimental data as well as the velocity V_ε.

Hence, instead of the Black–Scholes formula, we propose the following formula for the price X_t of a stock at time t:

$$X_t = X_0 \exp\left(x_0 + \varepsilon^{-1}\int_0^t v\left(\frac{s}{\varepsilon^2}\right) ds\right),$$

with the corresponding coefficients V_ε and Λ_ε.

4

Random Switched Processes with Delay in Reflecting Boundaries

In this chapter, we deal with the problem of finding the stationary distributions for the random switching processes in both a Markov and semi-Markov environment with delay in reflecting boundaries. These results can be used for the calculation of the stationary effectiveness parameter of multiphase supplying systems with feedback, as well as for determining the reliability of systems modeled by Markov and semi-Markov evolutions. Examples illustrating the application of these results are presented.

In section 4.1, we obtain the stationary distribution of the transport process with delaying boundaries in Markov media. In particular, these results are applicable to the study of multiphase systems with several buffers.

Section 4.2 deals with the transport process in semi-Markov switching. We find the stationary distribution of a random evolution, which is defined by the differential equation with the phase space on the interval $[V_0, V_1]$ and a constant vector field, whose values depend on the semi-Markov switching process with a finite set of states.

In section 4.3, we give a detailed example of the method of calculation for the stationary distributions of random switched processes, with delaying boundaries for the Markov case. In section 4.4, we provide the corresponding calculations for the semi-Markov switching process. By using these results, we calculate the stationary effectiveness parameters for single- and two-phase supplying systems with reservoirs and feedback.

4.1. Stationary distribution of evolutionary switched processes in a Markov environment with delay in reflecting boundaries

We will calculate the stationary distribution of an evolutionary switched process (or a transport process) with delaying barriers in a Markov environment. These results have particular application in the study of multiphase systems with several buffers (Pogorui and Turbin 2002; Pogorui 2004a).

Consider the following evolutionary switched process in a Markov environment:

$$\frac{dv(t)}{dt} = C\left(v(t), \kappa(t)\right) \tag{4.1}$$

where $\kappa(t)$ is a Markov process.

Let

$$G = X \cup Y, \qquad X = \{x_1, x_2, \ldots, x_n\}, \qquad Y = \{y_1, y_2, \ldots, y_m\}$$

be the phase space of $\kappa(t)$, and $P = \{p_{\alpha\beta},\ \alpha, \beta \in G\}$ be the transition probabilities matrix of the embedded (into $\kappa(t)$) ergodic Markov chain $\kappa_l,\ l \in \mathbb{N}$, i.e.

$$p_{\alpha\beta} = P\{\kappa_{l+1} = \beta |\ \kappa_l = \alpha\}.$$

A sojourn time τ_α of $\kappa(t)$ at $\alpha \in G$ is a random variable with exponential distribution $F_\alpha(t) = 1 - \exp\left(-\lambda_\alpha t\right)$.

Assume that the function C on the right-hand side of equation [4.1] is such that

C1. *For $x_i \in X$,*

$$C\left(v, x_i\right) = \begin{cases} -a_i, & V_0 < v \le V_1 \\ 0, & v = V_0 \end{cases}, \qquad i = 1, 2, \ldots, n,$$

for $y_j \in Y$,

$$C\left(v, y_j\right) = \begin{cases} b_j, & V_0 \le v < V_1 \\ 0, & v = V_1 \end{cases}, \qquad j = 1, 2, \ldots, m,$$

where V_0, V_1, a_i, $b_j \in \mathbb{R}$, $V_0 < V_1$, $a_i > 0$, $b_j > 0$, $i = 1, 2, \ldots, n$, $j = 1, 2, \ldots, m$.

We also assume the following:

C2. *The phase space G is composed of exactly one ergodic class (Korolyuk and Turbin 1982; Nummelin 1984).*

Thus, the process $\{v(t)\}$ develops in the segment (V_0, V_1) with constant velocities that are switched by the Markov process $\kappa(t)$. At point V_0 (respectively,

V_1), the process $v(t)$ delays up to the time when $\kappa(t)$ reaches the set Y (respectively, X).

We introduce a bivariate process $\{\xi(t)\}$ on the phase space $Z = G \times [V_0,\ V_1]$, i.e. $\xi(t) = (v(t), \kappa(t))$.

You should note that $\xi(t)$ is a Markov process (Korolyuk and Limnios 2005). Our aim is to study the stationary distribution of $\xi(t)$. The process $\xi(t)$ has an infinitesimal operator given by Korolyuk (1993); Korolyuk and Limnios (2005):

$$A\varphi\left(v,\alpha\right) = C\left(v,\alpha\right)\frac{d}{dv}\varphi\left(v,\alpha\right) + Q\varphi(v,\alpha),$$

where

$$Q\varphi\left(v,\alpha\right) = \lambda_\alpha[P\varphi\left(v,\alpha\right) - \varphi(v,\alpha)],$$

and

$$P\varphi\left(v,\alpha\right) = \sum_{\beta\in G} p_{\alpha\beta}\varphi\left(v,\beta\right).$$

If the process $\xi(t)$ has the stationary distribution ρ, then for any function $\varphi\left(\cdot\right)$ from the domain of A we have

$$\int_Z A\varphi\left(z\right)\rho\left(dz\right) = 0. \qquad [4.2]$$

Since the process $\xi(t)$ experiences delays at the points $\left(V_0, x\right),\ x\ \in\ X$ and $\left(V_1, y\right)\ y \in Y$, it follows that there are singularities (atoms) of the distribution ρ at these points. We will denote these singularities as $\rho\left[V_0,\ x\right]$, $\rho\left[V_1,\ y\right]$, respectively, and by $\rho\left(v,\alpha\right)$, $\alpha\ \in\ G$ as the absolute continuous components (or the probability density function) of the distribution ρ.

We assume that the following limits exist:

$$\rho\left(V_0^+,\,\alpha\right) = \lim_{v\downarrow V_0}\rho\left(v,\,\alpha\right), \qquad \rho\left(V_1^-,\,\alpha\right) = \lim_{v\uparrow V_1}\rho(v,\,\alpha).$$

Let A^* be the adjoint operator of A. By changing the order of integration in equation [4.2], we obtain the following expression for $A^*\rho = 0$, namely:

$$\int_Z A\varphi\left(z\right)\rho\left(dz\right) = \sum_{\alpha\in G}\int_{V_0}^{V_1}\left[C\left(v,\alpha\right)\frac{d}{dv}\varphi\left(v,\alpha\right)\right.$$

$$\left.+\lambda_\alpha\left(P\varphi\left(v,\alpha\right) - \varphi\left(v,\alpha\right)\right)\right]\rho\left(v,\alpha\right)dv$$

$$= \sum_{\alpha \in G} \Big(\varphi(V_1, \alpha) C(V_1, \alpha) \rho(V_1^-, \alpha) - \varphi(V_0, \alpha) C(V_0, \alpha) \rho(V_0^+, \alpha)$$

$$- \int_{V_0}^{V_1} \varphi(v, \alpha) C(v, \alpha) \frac{d}{dv} \rho(v, \alpha) \, dv \Big)$$

$$+ \sum_{\alpha \in G} \rho(v, \alpha) \lambda_\alpha (P\varphi(v, \alpha) - \varphi(v, \alpha)) = 0, \qquad [4.3]$$

where

$$\rho(V_0^+, \alpha) = \lim_{v \downarrow V_0} \rho(v, \alpha), \qquad \rho(V_1^-, \alpha) = \lim_{v \uparrow V_1} \rho(v, \alpha).$$

Thus, taking into account [4.2], for the pdf $\rho(\alpha, v)$ we have:

$$C(v, \alpha) \frac{d}{dv} \rho(v, \alpha) + Q^* \rho(v, \alpha) = 0, \quad \alpha \in G, \qquad V_0 < v < V_1, \qquad [4.4]$$

where Q^* is the adjoint operator of Q, i.e.

$$Q^* \rho(v, \alpha) = \sum_{\beta \in G} \lambda_\beta \, p_{\beta\alpha} \, \rho(v, \beta) + \lambda_\alpha \rho(v, \alpha).$$

Denoted by

$$\boldsymbol{\rho}[V_0] = (\rho[V_0, x_1], \ldots, \rho[V_0, x_n], 0, \ldots, 0),$$

$$\boldsymbol{\rho}[V_1] = (0, \ldots, 0, \rho[V_1, y_1], \ldots, \rho[V_1, y_m]),$$

$$\boldsymbol{\rho}(v) = (\rho(v, x_1), \ldots, \rho(v, x_n), \rho(v, y_1), \ldots, \rho(v, y_m))$$

vectors with respective $(n+m)$ components, and by

$$C = \operatorname{diag}(-a_1, -a_2, \ldots, -a_n, b_1, b_2, \ldots, b_m)$$

the $(n+m) \times (n+m)$ diagonal matrix.

The matrix representation of the operator Q is as follows:

$$Q = \begin{pmatrix} -\lambda_{x_1} & \lambda_{x_1} p_{x_1 x_2} & \cdots & \lambda_{x_1} p_{x_1 y_m} \\ \lambda_{x_2} p_{x_2 x_1} & -\lambda_{x_2} & \cdots & \lambda_{x_2} p_{x_2 y_m} \\ \vdots & \vdots & \ddots & \vdots \\ \lambda_{y_m} p_{y_m x_1} & \lambda_{y_m} p_{y_m x_2} & \cdots & -\lambda_{y_m} \end{pmatrix},$$

where we assume that $p_{\alpha\alpha} = 0$, $\alpha \in G$.

It follows from [4.3], that for the atoms of the distribution ρ, we have:

$$\boldsymbol{\rho}\left[V_0\right] Q=\boldsymbol{\rho}\left(V_0^{+}\right) C, \qquad \boldsymbol{\rho}\left[V_1\right] Q=\boldsymbol{\rho}\left(V_1^{-}\right) C. \qquad [4.5]$$

Let us consider equation [4.4] in the matrix form:

$$\frac{d}{dv}\boldsymbol{\rho}\left(v\right) C+\boldsymbol{\rho}\left(v\right) Q=\mathbf{0}, \qquad v \in\left(V_0, V_1\right). \qquad [4.6]$$

By solving equation [4.6], we have

$$\boldsymbol{\rho}\left(v\right)=\boldsymbol{\rho}\left(V_0^{+}\right) \exp \left(Q C^{-1}\left(V_0-v\right)\right), \qquad v \in\left(V_0,\, V_1\right). \qquad [4.7]$$

From [4.5] we can find the values of $\boldsymbol{\rho}\left(V_0^{+}\right)$ and the atoms $\boldsymbol{\rho}\left[V_0\right]$, $\boldsymbol{\rho}\left[V_1\right]$ up to a constant c. Indeed, by taking into account condition C2, the equation $Q\,\mathbf{x}=\mathbf{0}$ has the unique solution $\mathbf{x}=\mathbf{1}$, where $\mathbf{1}=(1, \ldots, 1)^{*}$ is a column vector with $(n+m)$ components. Therefore, formulas [4.5] consist of $2\,(n+m)-1$ independent equations, and the vectors $\boldsymbol{\rho}\left(V_0^{+}\right)$, $\boldsymbol{\rho}\left[V_0\right]$, $\boldsymbol{\rho}\left[V_1\right]$ contain $2\,(n+m)$ components.

It follows from [4.7], that the vector $\boldsymbol{\rho}\left(V_1^{-}\right)$ is uniquely expressed in terms of the vector $\boldsymbol{\rho}\left(V_0^{+}\right)$. A normalization constant c is determined by the following condition:

$$\int_{V_0}^{V_1} \boldsymbol{\rho}\left(v\right) \mathbf{1}\, dv+\left(\boldsymbol{\rho}\left[V_0\right]+\boldsymbol{\rho}\left[V_1\right]\right) \mathbf{1}=1. \qquad [4.8]$$

Thus, we have the following theorem:

THEOREM 4.1.– (Pogorui 2005) If conditions C1 and C2 are satisfied, then the Markov process $\xi(t)=(v(t), \kappa(t))$ has the stationary distribution $\boldsymbol{\rho}\left(\cdot\right)$, with the absolutely continuous part $\boldsymbol{\rho}\left(v\right)=\boldsymbol{\rho}\left(V_0^{+}\right) \exp \left(Q C^{-1}\left(V_0-v\right)\right)$, $v \in\left(V_0, V_1\right)$ and singular part (atoms) $\boldsymbol{\rho}\left[V_0\right]$, $\boldsymbol{\rho}\left[V_1\right]$ determined by [4.5].

EXAMPLE 4.1.– Let us consider the case where $X=\{x\}$, $Y=\left\{y_1, y_2, \ldots, y_m\right\}$. Then $\boldsymbol{\rho}\left[V_0\right]=\left(\rho\left[V_0, x\right], 0, \ldots, 0\right)$, $\boldsymbol{\rho}\left[V_1\right]=\left(0,\, \rho\left[V_1, y_1\right], \ldots,\, \rho\left[V_1, y_m\right]\right)$ and the matrices C and Q are of the following form:

$$C=\begin{pmatrix} -a & 0 & \ldots & 0 \\ 0 & b_1 & \ldots & 0 \\ \vdots & \vdots & \ddots & \vdots \\ 0 & 0 & \ldots & b_m \end{pmatrix},$$

$$Q=\begin{pmatrix} -\lambda_x & \lambda_x p_{x y_1} & \ldots & \lambda_x p_{x y_m} \\ \lambda_{y_1} p_{y_1 x} & -\lambda_{y_1} & \ldots & \lambda_{y_1} p_{y_1 y_m} \\ \vdots & \vdots & \ddots & \vdots \\ \lambda_{y_m} p_{y_m x} & \lambda_{y_m} p_{y_m y_1} & \ldots & -\lambda_{y_m} \end{pmatrix}.$$

From equation [4.5], it follows that

$$\rho\left(V_0^+, x\right) = \frac{\lambda_x \rho\left[V_0, x\right]}{a}, \qquad \rho\left(y_1, V_0^+\right) = \frac{\lambda_x p_{xy_1} \rho\left[V_0, x\right]}{b_1},$$

$$\dots, \qquad \rho\left(y_m, V_0^+\right) = \frac{\lambda_x p_{xy_m} \rho\left[V_0, x\right]}{b_m}.$$

and

$$\begin{aligned}&(0, \rho\left[V_1, y_1\right], \dots, \rho\left[V_1, y_m\right]) Q\\&= \boldsymbol{\rho}\left(V_1^-\right) C = \boldsymbol{\rho}\left(V_0 +\right) \exp\left(-QC^{-1}\left(V_1 - V_0\right)\right) C.\end{aligned} \quad [4.9]$$

Denote as Q_Y and C_Y, two matrices obtained from the matrices Q and C respectively, after removing the first column and the first row. Consider

$$\boldsymbol{\rho}_Y\left(V_1^-\right) = \left(\rho\left(V_1^-, y_1\right), \dots, \rho(V_1^-, y_m)\right).$$

It follows from condition C2, that an inverse matrix Q_Y^{-1} exists. Thus, it is easily verified that

$$\begin{aligned}&\left(\rho\left[V_1, y_1\right], \dots, \rho\left[V_1, y_m\right]\right)\\&\quad = \left(\rho\left(V_1^+, y_1\right), \dots, \rho\left(V_1^+, y_{m'}\right)\right) C_Y Q_Y^{-1}.\end{aligned} \quad [4.10]$$

Since the vector $\boldsymbol{\rho}_Y\left(V_1^-\right)$ is uniquely expressed in terms of the vector $\boldsymbol{\rho}\left(V_0^+\right)$, then up to the constant $c = \rho\left[V_0, x\right]$, we have the absolutely continuous part of the stationary distribution

$$\boldsymbol{\rho}\left(v\right) = \left(\frac{\lambda_x c}{a}, \frac{\lambda_{y_1} c}{b_1}, \dots, \frac{\lambda_{y_m} c}{b_m}\right) \exp(QC^{-1}(V_0 - v)).$$

The atoms of the stationary distribution are determined by equation [4.10] and the constant $c = \rho\left[x, V_0\right]$ can be obtained from equation [4.8].

For the estimation of efficiency of some storage systems, it is enough to calculate the value of $\rho_s\left(v\right) = \sum_{i=1}^m \rho\left(v, x_i\right)$. In order to obtain $\rho_s\left(v\right)$, let us consider the following differential operator:

$$C\left(v, \alpha\right) \frac{d}{dv} + Q^*$$

$$= \begin{pmatrix} -\lambda_{x_1} - a_1 \frac{d}{dv} & \lambda_{x_2} p_{x_2 x_1} & \dots & \lambda_{y_m} p_{y_m x_1} \\ \lambda_{x_1} p_{x_1 x_2} & -\lambda_{x_2} - a_2 \frac{d}{dv} & \dots & \lambda_{y_m} p_{y_m x_2} \\ \vdots & \vdots & \ddots & \vdots \\ \lambda_{x_1} p_{x_1 y_m} & \lambda_{x_2} p_{x_2 y_m} & \dots & -\lambda_{y_m} + b_m \frac{d}{dv} \end{pmatrix} \quad [4.11]$$

Let the differential operator D be defined as a formal determinant of the operator $C(v,\alpha)\frac{d}{dv}+Q^*$, i.e. $D=\det\left(C(v,\alpha)\frac{d}{dv}+Q^*\right)$. As it was mentioned in Volume 1, Chapter 3 (see the proof of lemma 3.4), the function $\rho_s(v)$ satisfies the following differential equation of the order $n+m$:

$$D\rho_s(v)=0.$$

EXAMPLE 4.2.– Let us consider a similar case, such as that of example 4.1, where $m=2$, i.e. $X=\{x\}, Y=\{y_1, y_2\}$. Then,

$$C(v,\alpha)\frac{d}{dv}+Q^*=\begin{pmatrix} -\lambda_x-a\frac{d}{dv} & \lambda_{y_1}p_{y_1x} & \lambda_{y_2}p_{y_2x} \\ \lambda_x p_{xy_1} & -\lambda_{y_1}+b_1\frac{d}{dv} & \lambda_{y_2}p_{y_2y_1} \\ \lambda_x p_{xy_2} & \lambda_{y_1}p_{y_1y_2} & -\lambda_{y_2}+b_2\frac{d}{dv}\end{pmatrix}.$$

Hence,

$$\begin{aligned} D\rho_s(v) &= \det\left(C(v,\alpha)\frac{d}{dv}+Q^*\right)\rho_s(v) \\ &= -ab_1b_2\frac{d^3}{dv^3}\rho_s(v)+\left(-\lambda_x b_1b_2+\lambda_{y_1}ab_2+\lambda_{y_1}ab_1\right)\frac{d^2}{dv^2}\rho_s(v) \\ &+\left(\lambda_x\lambda_{y_1}b_2\left(1-p_{xy_1}p_{y_1x}\right)+\lambda_x\lambda_{y_2}b_1\left(1-p_{xy_2}p_{y_2x}\right)\right. \\ &\quad \left.-\lambda_{y_1}\lambda_{y_2}a\left(1-p_{y_1y_2}p_{y_2y_1}\right)\right)\frac{d}{dv}\rho_s(v) \\ &+\lambda_x\lambda_{y_1}\lambda_{y_2}\left(p_{y_1y_2}p_{y_2y_1}+p_{xy_1}p_{y_1x}+p_{xy_2}p_{y_2x}+p_{xy_1}p_{y_1y_2}p_{y_2x}\right. \\ &\quad \left.+p_{xy_2}p_{y_2y_1}p_{y_1x}-1\right)\rho_s(v)=0. \end{aligned}$$

4.2. Stationary distribution of switched process in semi-Markov media with delay in reflecting barriers

This section deals with a switched process that has semi-Markov switching (or, in other words, a switched process that is in a semi-Markov media) (Korolyuk and Limnios 2005). We find the stationary measure of this process, which is defined by the differential equation with the phase space on the interval $[V_0, V_1]$ and a constant vector field of values that depends on the semi-Markov switching process, with a finite set of states.

The results obtained in this section can be used in reliability theory for the calculation of the stationary effectiveness of systems (Pogorui 2011b; Rodríguez-Said *et al.* 2007, 2008). In particular, it can be used in the study of multiphase systems with storage that can be modeled by stochastic transport processes, with delay in barriers with a Markov or semi-Markov switching process

(Pogorui and Turbin 2002; Pogorui 2004a). If the switching process is semi-Markov, then we can obtain the stationary distribution of the corresponding three-variate Markov process, in which the first temporal component (i.e. the time that is passed after the last change of a state by the switching process), the second component corresponds to the switching process and the third one is the spatial component describing the fullness of the storage facility. We should remark that the problem of finding the stationary distribution of the switched process in semi-Markov media is a nontrivial problem, even for the simplest case of an alternative switching process (Pogorui 2011b).

4.2.1. *Infinitesimal operator of random evolution with semi-Markov switching*

We will study the infinitesimal operator of a stochastic switched process in a random environment, which can be seen as the superposition of independent semi-Markov processes. Such a process is defined by the evolutionary differential equation, which depends on the given superposition. These types of processes are applied to the calculation of the efficiency of various feedback control systems (Pogorui and Turbin 2002; Pogorui 2003), and in some queueing systems (Foss 2007). Another example of the application of such processes is considered in Volume 2, Chapter 5, where the distribution of the first crossing time of two independent telegraph processes on the line, is studied.

Let us consider the evolutionary switched process $\{V(t)\}$, which is defined as a solution of the following evolutionary differential equation:

$$\frac{dV(t)}{dt} = C\left(V(t), \xi_1(t), \xi_2(t), \dots, \xi_n(t)\right), \qquad V(0) = v, \qquad [4.12]$$

where $\xi_i(t)$, $i = 1, 2, \dots, n$, is a family of independent regular semi-Markov processes with the respective phase spaces $X_i = \left\{x_1^{(i)}, x_2^{(i)}, \dots, x_{m_i}^{(i)}\right\}$ and semi-Markov matrices $Q^{(i)}(t) = \left[p_{x_k^{(i)} x_l^{(i)}} G_{x_k^{(i)}}(t);\, 1 \leq k,\, l \leq m_i\right]$, and the function $C\left(v, x_1, x_2, \dots, x_n\right)$, $x_i \in X_i$, is continuously differentiable on $\mathbb{R} \times \prod_{i=1}^{n} X_i$ with respect to v for all $x_i \in X_i$, $i = 1, 2, \dots, n$.

Denote by p_{yz}, $y, z \in X_i$ the transition probabilities of the embedded (in $\xi_i(t)$) Markov chain $\xi_n^{(i)} = \xi_i\left(\tau_n^{(i)}\right)$, where $\tau_n^{(i)}$, $n = 0, 1, 2, \dots$, are renewal epochs of the renewal process $\xi_i(t)$, and $G_{x_k^{(i)}}(t)$ is the sojourn time distribution function of the process $\xi_i(t)$ at $x_k^{(i)}$.

Let us assume that the pdf $g_{x_k^{(i)}}(t) = \frac{dG_{x_k^{(i)}}(t)}{dt}$ exists for all $x_{k_i}^{(i)} \in X_i$, $i = 1, 2, \ldots, n$, and define

$$r_{x_k^{(i)}}(\tau_i) = \frac{g_{x_k^{(i)}}(t)}{1 - G_{x_k^{(i)}}(t)}.$$

Now, let us consider $\tau_i(t) = t - \inf\{u \leq t : \xi_i(u) \neq \xi_i(t)\}$, $i = 1, 2, \ldots, n$. Since the bivariate process $(\xi_i(t), \tau_i(t))$ is a Markov process (Corlat *et al.* 1991; Gikhman and Skorokhod 1975; Korolyuk and Turbin 1982), it is easy to verify that the process

$$\zeta(t) = (V(t),\ \xi_1(t),\ \tau_1(t),\ \xi_2(t),\ \tau_2(t), \ldots,\ \xi_n(t),\ \tau_n(t))$$

is a Markov process as well.

LEMMA 4.1.– The infinitesimal operator of the process $\zeta(t)$ is as follows:

$$Gf(v, x_1, \tau_1, x_2, \tau_2, \ldots, x_n, \tau_n)$$

$$= C(v, x_1, \tau_1, x_2, \tau_2, \ldots, x_n, \tau_n) \frac{\partial}{\partial v} f(v, x_1, \tau_1, x_2, \tau_2, \ldots, x_n, \tau_n)$$

$$+ \sum_{i=1}^{n} \frac{\partial f(v, x_1, \tau_1, x_2, \tau_2, \ldots, x_n, \tau_n)}{\partial \tau_i}$$

$$+ r_{x_1}(\tau_1)(P_1 f(v, x_1, 0, x_2, \tau_2, \ldots, x_n, \tau_n) - f(v, x_1, \tau_1, x_2, \tau_2, \ldots, x_n, \tau_n))$$

$$+ r_{x_2}(\tau_2)(P_2 f(v, x_1, \tau_1, x_2, 0, \ldots, x_n, \tau_n) - f(v, x_1, \tau_1, x_2, \tau_2, \ldots, x_n, \tau_n)),$$

$$+ \ldots$$

$$+ r_{x_n}(\tau_n)(P_n f(v, x_1, \tau_1, x_2, \tau_2, \ldots, x_n, 0) - f(v, x_1, \tau_1, x_2, \tau_2, \ldots, x_n, \tau_n)),$$

where *f is in the domain of the operator G.*

$$P_i f(v, x_1, \tau_1, \ldots, x_i, 0, \ldots, x_n, \tau_n)$$

$$= \sum_{y \in E_i} p_{x_i y} f(v, x_1, \tau_1, \ldots, y, 0, \ldots, x_n, \tau_n).$$

REMARK 4.1.– *Let us consider a particular case*

$$\frac{dV(t)}{dt} = C(V(t), \xi(t)), \qquad V(0) = v,$$

that is, the "superposition" consists of one semi-Markov process $\xi(t)$.

Let us state $\tau(t) = t - \sup\{u : \xi(u) \neq \xi(t)\}$.

On the phase space, $\mathbb{R}^n \times X \times [0, +\infty)$ *introduce the three-variate process* $\zeta(t) = (V(t), \xi(t), \tau(t))$, *where* $\tau(t) = t - \sup\{u \leq t : \xi(u) \neq \xi(t)\}$.

Then, $\zeta(t)$ *is a Markov process and its infinitesimal operator* G *is given by*

$$Gf(v, x, t) = C(v, x) \frac{\partial}{\partial v} f(v, x, t)$$

$$+r_x(t) [Pf(v, x, 0) - f(v, x, t)] + \frac{\partial}{\partial t} f(v, x, t),$$

where $f(v, x, \tau)$ *is a bounded and continuously differentiable function with respect to* v *and* τ.

PROOF.– Consider the proof of lemma 4.1 for the case of one switching process, considered in remark 4.1.

Suppose $\xi(0) = x$, $\tau(0) = t$, $\xi_n(0) = x^{(n)}_{k_n}$, $\tau_n(0) = t_n$.

Taking into account the definition of the infinitesimal operator of a Markov process (Gikhman and Skorokhod 1975), we have

$$Gf(v, x, t) = \lim_{\Delta \to 0} \frac{\mathbf{E}_{v,x,t}[f(V(\Delta), \xi(\Delta), \tau(\Delta))] - f(v, x, t)}{\Delta}.$$

To simplify expressions involving $\mathbf{E}_{v,x,t}$, the subscript (v, x, t) will be omitted. We present this expectation in the following form:

$$\mathbf{E}[f(V(\Delta), \xi(\Delta), \tau(\Delta))] = f(v, x, t)$$

$$+\mathbf{E}[f(V(\Delta), \xi(\Delta), \tau(\Delta)) - f(v, \xi(\Delta), \tau(\Delta))]$$

$$+\mathbf{E}[f(v, \xi(\Delta), \tau(\Delta)) - f(v, x, t)].$$

Denoting a sojourn time of $\xi(t)$ at the state x as θ_x, we obtain

$$\mathbf{E}[f(V(\Delta), \xi(\Delta), \tau(\Delta)) - f(v, x, t)]$$

$$= \mathbf{E}[f(V(\Delta), \xi(\Delta), \tau(\Delta)) - f(v, \xi(\Delta), \tau(\Delta))] \mathbb{I}(\theta_x > \Delta)$$

$$+\mathbf{E}[f(V(\Delta), \xi(\Delta), \tau(\Delta)) - f(v, \xi(\Delta), \tau(\Delta))] \mathbb{I}(\theta_x \leq \Delta)$$

$$+\mathbf{E}[f(v, \xi(\Delta), \tau(\Delta)) - f(v, x, t)].$$

Since the semi-Markov process $\xi(t)$ is regular and f is differentiable with respect to v, we have

$$\mathbf{E}[f(V(\Delta), \xi(\Delta), \tau(\Delta)) - f(v, \xi(\Delta), \tau(\Delta))] \mathbb{I}(\theta_x \leq \Delta) = o(\Delta).$$

Then, using the fact that $V(\Delta) = v + \int_0^{\Delta} C(V(t), \xi(t))dt$, for differentiable function f with respect to v, we obtain

$$\lim_{\Delta \to 0} \Delta^{-1}\mathbf{E}\left[f\left(V\left(\Delta\right), \xi\left(\Delta\right), \tau\left(\Delta\right)\right) - f\left(v, \xi\left(\Delta\right), \tau\left(\Delta\right)\right)\right]\mathbb{I}\left(\theta_x > \Delta\right)$$

$$= \lim_{\Delta \to 0} \Delta^{-1}\mathbf{E}\left[f\left(v + \int_0^{\Delta} C\left(V(t), \xi(t)\right)dt, \xi\left(\Delta\right), \tau\left(\Delta\right)\right)\right.$$

$$\left. - f\left(v, \xi\left(\Delta\right), \tau\left(\Delta\right)\right)\right]\mathbb{I}\left(\theta_x > \Delta\right) = C\left(v, x\right)\frac{\partial}{\partial v}f\left(v, x, t\right). \qquad [4.13]$$

It is well known that the bivariate process $(\xi(t), \tau(t))$ is a Markov process and its infinitesimal operator is of the following form (Volume 1, Chapter 1, section 1.4) (Corlat *et al.* 1991; Korolyuk and Turbin 1982):

$$Ah\left(x, t\right) = \lim_{\Delta \to 0} \frac{\mathbf{E}_{x,t}\left[h\left(\xi\left(\Delta\right), \tau\left(\Delta\right)\right)\right] - h\left(x, t\right)}{\Delta}$$

$$= r_x(t)\left[Ph\left(x, 0\right) - h\left(x, t\right)\right] + \frac{\partial}{\partial t}h\left(x, t\right),$$

where $h\left(x, t\right)$ is bounded and differentiable with respect to t.

Hence,

$$\lim_{\Delta \to 0} \frac{\mathbf{E}\left[f\left(v, \xi\left(\Delta\right), \tau\left(\Delta\right)\right) - f\left(v, x, t\right)\right]}{\Delta}$$

$$= r_x(t)\left[Pf\left(v, x, 0\right) - f\left(v, x, t\right)\right] + \frac{\partial}{\partial t}f(v, x, t). \qquad [4.14]$$

Taking into account [4.13] and [4.14], we obtain the proof of lemma 4.1 for one switching process. For the general case, the proof follows the same idea.

4.2.2. *Stationary distribution of random evolution in semi-Markov media with delaying boundaries in balance case*

In this section, we study a switched process with semi-Markov switching, that is defined by an evolution differential equation with a finite vector field. The phase space of the differential equation is a segment of the real line, and its vector field consists of constants. We obtain the stationary distribution of the switched process given by the evolution differential equation.

Let $\{\kappa(t)\}$ be a semi-Markov process with phase space $G = X \cup Y$, $X = \{x_1, x_2, \ldots, x_n\}$, $Y = \{y_1, y_2, \ldots, y_m\}$, and $P = \{p_{\alpha\beta}, \alpha, \beta \in G\}$ as the transition probabilities matrix of the embedded (into $\kappa(t)$) Markov chain κ_l, $l \in \mathbb{N}$, i.e. $p_{\alpha\beta} = P\{\kappa_{l+1} = \beta \mid \kappa_l = \alpha\}$.

We denote by τ_α a sojourn time of $\kappa(t)$ at $\alpha \in G$, and by $F_\alpha(t)$, the cumulative distribution function of τ_α.

Consider the evolutionary switched process $v(t)$ defined as a solution of the following evolutionary equation:

$$\frac{dv(t)}{dt} = C(\kappa(t), v(t)). \qquad [4.15]$$

The process $v(t)$ is also called a random evolution (or a transit process) in a semi-Markov media $\kappa(t)$ (Korolyuk 1993; Korolyuk and Swishchuk 1995b; Korolyuk and Limnios 2005).

Now, let us introduce the following condition:

D$_1$. *The density function* $f_\alpha(t) = \frac{dF_\alpha(t)}{dt}$, *and moments*

$$m_\alpha = \int_0^\infty t f_\alpha(t)dt, \quad m_\alpha^{(2)} = \int_0^\infty t^2 f_\alpha(t)dt \quad \forall \alpha \in G \; exist.$$

Consider $V_0, V_1, a_i, b_j \in \mathbb{R}$, $V_0 < V_1$, $a_i > 0$, $b_j > 0$, $i = 1, 2, \dots, n$, $j = 1, 2, \dots, m$, and assume that the function C, of the right-hand side of equation [4.15], satisfies the following conditions:

$$C(x_i, v) = \begin{cases} b_i, & V_0 \le v < V_1, \\ 0, & v = V_1, \end{cases} \qquad \text{for } x_i \in X, \; i = 1, 2, \dots, n$$

$$C(y_j, v) = \begin{cases} -a_j, & V_0 < v \le V_1, \\ 0, & v = V_0. \end{cases} \qquad \text{for } y_j \in Y, \; j = 1, 2, \dots, m.$$

On the phase space $Z = [0, \infty) \times G \times [V_0, V_1]$, consider the three-variate process $\xi(t) = (\tau(t), \kappa(t), v(t))$, where $\tau(t) = t - \sup\{u \le t : \kappa(t) \ne \kappa(u)\}$.

Our aim is to calculate the stationary distribution of $\xi(t)$.

The process $\xi(t)$ is a Markov process and its infinitesimal operator is of the following form:

$$A\varphi(\tau, \alpha, v) = \frac{\partial}{\partial \tau}\varphi(\tau, \alpha, v) + r_\alpha(\tau)\left[P\varphi(0, \alpha, v) - \varphi(\tau, \alpha, v)\right]$$

$$+ C(\alpha, v)\frac{\partial}{\partial v}\varphi(\tau, \alpha, v),$$

with the boundary conditions $\varphi'_\tau(\tau, x, V_0) = \varphi'_\tau(\tau, y, V_1) = 0$, $x \in X$, $y \in Y$, where

$$r_\alpha(\tau) = \frac{f_\alpha(\tau)}{1 - F_\alpha(\tau)}, \qquad P\varphi(0,\alpha, v) = \sum_{\beta \in G} p_{\alpha\beta}\, \varphi(0, \beta, v).$$

If the process $\xi(t)$ has the stationary distribution $\rho(\cdot)$, then for any function $\varphi(\cdot)$, from the domain of A, we have:

$$\int_Z A\varphi(z)\, \rho(dz) = 0. \qquad [4.16]$$

Consider the following condition:

D$_2$. *The pdf $\rho(\tau, \alpha, v)$, of the absolutely continuous part of the stationary distribution of the process $\xi(t)$, exists. In addition, everywhere, $\frac{\partial \rho(\tau,\alpha,v)}{\partial \tau}$, $\frac{\partial \rho(\tau,\alpha,v)}{\partial v}$, and the following limits* $\lim_{\tau \to +\infty} \rho(\tau, \alpha, v)$, $\lim_{v \uparrow V_1} \rho(\tau, x, v)$, *and* $\lim_{v \downarrow V_0} \rho(\tau, x, v)$ exist.

The analysis of the properties of the process $\xi(t)$ leads to the conclusion that the stationary distribution $\rho(\cdot)$ has singularities at points (τ, x, V_1), $x \in X$ and (τ, y, V_0), $y \in Y$, which are denoted as $\rho[\tau, x, V_0]$, $\rho[\tau, y, V_1]$.

Now, let A^* be the adjoint operator of A. Then, by changing the order of integration in equation [4.16] (integrating by parts), we can obtain the following expressions for the non-singular part of $A^*\rho - 0$:

$$C(\alpha, v) \frac{\partial}{\partial v} \rho(\tau, \alpha, v) + r_\alpha(\tau)\, \rho(\tau, \alpha, v) + \frac{\partial}{\partial \tau} \rho(\tau, \alpha, v) = 0, \qquad [4.17]$$

$$\rho(\infty, \alpha, v) = 0, \qquad \alpha \in G,$$

$$\sum_{\beta \in G} \int_0^\infty r_\beta(\tau)\, \rho(\tau, \beta, v)\, d\tau, \qquad p_{\beta\alpha} = \rho(0,\alpha, v), \quad \alpha \in G. \qquad [4.18]$$

And for the singular part we have:

$$\rho[\infty, x, V_1] = 0, \quad x \in X, \qquad \rho[\infty, y, V_0] = 0, \quad y \in Y$$

$$\frac{d}{d\tau} \rho[\tau, x, V_1] + r_x(\tau)\, \rho[\tau, x, V_1] - b_x \rho\left(\tau, x, V_1^-\right) = 0, \qquad [4.19]$$

$$\frac{d}{d\tau} \rho[\tau, y, V_0] + r_y(\tau)\, \rho[\tau, y, V_0] - a_y \rho\left(\tau, y, V_0^+\right) = 0, \qquad [4.20]$$

where

$$\rho\left(\tau, x, V_1^-\right) = \lim_{v \uparrow V_1} \rho(\tau, x, v), \qquad \rho\left(\tau, x, V_0^+\right) = \lim_{v \downarrow V_0} \rho(\tau, x, v).$$

Then,

$$\sum_{y\in Y}\int_0^{\infty} r_y(\tau)\,\rho[\tau,y,V_0]\,d\tau \qquad p_{yz}=\rho[0,z,V_0]\,,\ z\in Y,$$

$$\sum_{x\in X}\int_0^{\infty} r_x(\tau)\,\rho[\tau,x,V_0]\,d\tau \qquad p_{xz}=\rho[0,z,V_0]\,,\ z\in X, \qquad [4.21]$$

$$\sum_{y\in Y}\int_0^{\infty} r_y(\tau)\,\rho[\tau,y,V_0]\,d\tau \quad p_{yx}=b_x\int_0^{\infty}\rho\left(\tau,x,V_0^{+}\right)d\tau,\ x\in X,$$

$$\sum_{x\in X}\int_0^{\infty} r_x(\tau)\,\rho[\tau,x,V_1]\,d\tau \quad p_{xy}=a_y\int_0^{\infty}\rho\left(\tau,y,V_1^{-}\right)d\tau,\ y\in Y.$$

By solving equation [4.17], we have:

$$\rho(\tau,x,v)=f_x(v-b_x\tau)\ e^{-\int_0^{\tau} r_x(s)ds},\ x\in X,$$

$$\rho(\tau,y,v)=f_y(v+a_y\tau)\ e^{-\int_0^{\tau} r_y(s)ds},\ y\in Y.$$

The balance case is characterized by the independence of the non-singular part, from the stationary distribution $\rho(\cdot)$ on v.

In this case, functions $f_x(v-b_x\tau)=c_x$, $f_y(v+a_y\tau)=c_y$ are constants. Hence, taking into account that $e^{-\int_0^{\tau} r_\alpha(s)ds}=1-F_\alpha(\tau)$, we have the following solution to equation [4.17]:

$$\rho(\tau,\alpha,v)=c_\alpha\,(1-F_\alpha(\tau))\,, \qquad \alpha\in G. \qquad [4.22]$$

If condition D_1 is fulfilled, then $\frac{\partial\rho(\tau,\alpha,v)}{\partial\tau}=-c_\alpha f_\alpha(\tau)$, $\lim_{\tau\to+\infty}(\tau,\alpha,v)=0$.

Since $\rho(\tau,\alpha,v)$ does not depend on v, it is easily verified that the solution [4.22] satisfies the condition C_2.

By substituting $\rho(\tau,\alpha,v)=c_\alpha(1-F_\alpha(\tau))$ into [4.18], we have

$$\sum_{\alpha\in G} c_\alpha p_{\alpha\beta}=c_\beta. \qquad [4.23]$$

We suppose that the following condition holds:

D_3. *The embedded (into $\kappa(t)$) Markov chain is irreducible, ergodic and has the stationary distribution ρ_α, $\alpha\in G$.*

Then a solution of [4.23] is as follows:

$$c_\alpha = c\rho_\alpha, \qquad \alpha \in G. \tag{4.24}$$

Hence,

$$\int_0^\infty \rho(\tau, \alpha, v)\, d\tau = c\rho_\alpha \int_0^\infty (1 - F_\alpha(\tau))\, d\tau = c\rho_\alpha m_\alpha, \quad \alpha \in G. \tag{4.25}$$

By solving [4.19], [4.20], and considering [4.22] and [4.24], we obtain

$$\begin{aligned} \rho[\tau,\, x, V_1] &= c\,\rho_x (1 - F_x(\tau)) \left(b_x \tau + \frac{\rho[0, x, V_1]}{\sigma \rho_x} \right), \\ \rho[\tau,\, y, V_0] &= c\,\rho_y (1 - F_y(\tau)) \left(a_y \tau + \frac{\rho[0, y, V_0]}{\sigma \rho_y} \right). \end{aligned} \tag{4.26}$$

Substituting [4.26] into [4.21], we have

$$\begin{aligned} &\sum_{x\in X} \rho_x b_x m_x p_{xz} + \sum_{x\in X} \rho[0, x, V_1]\, p_{xz} = \rho[0, z, V_1], \quad z \in X, \\ &\sum_{y\in Y} \rho_y a_y m_y p_{yz} + \sum_{y\in Y} \rho[0, y, V_0]\, p_{yz} = \rho[0, y, V_0], \quad z \in Y, \\ &\sum_{y\in Y} \rho_y a_y\, m_y p_{yx} + \sum_{y\in Y} \rho[0, y, V_0]\, p_{yx} = b_x m_x \rho_x, \quad x \in X, \\ &\sum_{x\in X} \rho_x b_x m_x p_{xy} + \sum_{x\in X} \rho[0, x, V_1]\, p_{xy} = a_y m_y \rho_y, \quad y \in Y. \end{aligned} \tag{4.27}$$

Denote by $P_X = \{p_{xz},\ x, z \in X\}$, $P_Y = \{p_{yz},\ y, z \in X\}$, $G_X = (I - P_X)^{-1} = \{g_{xz}, x,\ z \in X\}$, $G_Y = (I - P_Y)^{-1} = \{g_{yz}, y,\ z \in Y\}$. The matrices G_X, G_Y resemble the potential, as it is described in Kemeny and Snell (1960); Korolyuk and Turbin (1993).

By solving the first two equations of [4.21], we obtain

$$\begin{aligned} \rho[0, z, V_0] &= \sum_{x\in X} \rho_x b_x m_x \sum_{k\in X} p_{xk} g_{kz}, \qquad z \in X, \\ \rho[0, z, V_1] &= \sum_{y\in Y} \rho_y a_y m_y \sum_{k\in Y} p_{yk} g_{kz}, \qquad z \in Y. \end{aligned}$$

After the substitution of these relations into the last two equations of [4.21], we obtain the following balance condition:

$$\text{D}_4. \sum_{x\in X} b_x m_x \rho_x \, p_{xy} + \sum_{x\in X} \rho_x b_x m_x \sum_{k\in X} p_{xk} \sum_{z\in X} g_{kz} p_{zy} = b_y m_y \rho_y, \; y \in Y,$$

$$\sum_{y\in Y} a_y m_y \rho_y \, p_{yx} + \sum_{y\in Y} \rho_y a_y m_y \sum_{k\in Y} p_{yk} \sum_{z\in Y} g_{kz} p_{zx} = b_x m_x \rho_x, \; x \in X. \quad [4.28]$$

Thus, we have proved the following theorem:

THEOREM 4.2.– If conditions $\text{D}_1 - \text{D}_4$ are satisfied, then the stationary distribution $\rho(\cdot)$ of $\xi(t)$ has the following form:

for the continuous part ($x \in X$ and $y \in Y$)

$$\rho(\tau, x, v) = c\rho_x \; (1 - F_x(\tau)); \qquad \rho(\tau, y, v) = c\rho_y \, (1 - F_y(\tau)),$$

and for the singular part (atoms)

$$\rho[\tau, x, V_0] = c\rho_x \; (1 - F_x(\tau)) \left(b_x \tau + \frac{\rho[0, x, V_0]}{c\rho_x} \right),$$

$$\rho[\tau, y, V_1] = c\rho_y \; (1 - F_y(\tau)) \left(a_y \tau + \frac{\rho[0, y, V_1]}{c\rho_y} \right),$$

where

$$\rho[0, z, V_0] = \sum_{x\in X} \rho_x b_x \, m_x \sum_{k\in X} p_{xk} g_{kz}, \quad z \in X;$$

$$\rho[0, z, V_1] = \sum_{y\in Y} \rho_y a_y \, m_y \sum_{k\in Y} p_{yk} g_{kz}, \quad z \in Y.$$

The constant c is determined from the condition

$$\int_Z \rho(dz) = 1.$$

Now, let us show that ρ of theorem 4.2 is a unique measure, i.e. there is no other stationary distribution that satisfies equation [4.16], even if condition D_2 is not fulfilled.

As we know, the coupling method is well known in the theory of random processes. The method is widely used in the investigation of Markov processes and similar processes (Lindvall 2002; Yu. 2002). The following lemma is true:

LEMMA 4.2.– (Yu. 2002) Suppose that a process $\zeta(t)$, with phase space Ψ, satisfies the following condition:

(C) $T > 0$, $\gamma > 0$ and the two-dimensional vector Markov process $\boldsymbol{\theta}(t) = (\theta_1(t), \theta_2(t))$ exist on the phase space Ψ^2, for which the following assertions are true:

1) both components of the process $\boldsymbol{\theta}(t)$ have the same distributions as the process $\zeta(t)$;

2) for any x and y, one has $P\{\theta_1(t) = \theta_2(t) \,|\, \theta_1(0) = x, \theta_2(0) = y\} \geq \gamma$.

Then the process $\zeta(t)$ has, at most, one stationary distribution.

Let us show how the coupling method can be used in the present case.

I. Let $\kappa(t)$ be a Markov process, i.e. instead of the three-variate process $\xi(t) = (\tau(t), \kappa(t), \upsilon(t))$, we can consider the bivariate process $\overline{\xi}(t) = (\kappa(t), \upsilon(t))$.

Without the loss of generality, we assume that the embedded (into $\kappa(t)$) Markov chain $\{\kappa_l,\ l \geq 1\}$ satisfies the following condition:

$$p_{\alpha\beta} = P\{\kappa_{l+1} = \alpha \,|\, \kappa_l = \beta\} \geq \delta > 0, \quad \forall \alpha, \beta \in G.$$

Let $\kappa_l{}'$, $l \geq 1$, and $\kappa_l{}''$, $l \geq 1$, be two independent Markov chains with common phase space G, transition-probability matrix $P = \{p_{\alpha\beta},\ \alpha, \beta \in G\}$ and simultaneous jumps. It is easy to verify that

$$P\{\kappa_1{}' = \alpha_0,\ \kappa_1{}'' = \alpha_0 \,|\, \kappa_0{}' = \alpha,\ \kappa_0{}'' = \beta\} \geq \delta^2 > 0,\ \ \forall \alpha_0,\ \alpha, \beta \in G.$$

Thus, after the first jump, the processes $\kappa_l{}'$, $l \geq 1$, and $\kappa_l{}''$, $l \geq 1$, are coupled with a probability not less than $\delta^2 > 0$. Now, consider two independent Markov processes $\kappa'(t)$ and $\kappa''(t)$ with embedded chains $\kappa_l{}'$ and $\kappa_l{}''$, respectively, and the same sojourn intensities as for $\kappa(t)$. Furthermore, in the period of time preceding a certain epoch $T_0 > 0$, the process $\kappa(t)$ jumps with a probability not less than $\delta_1 = \lambda \int_0^{T_0} e^{-\lambda s} ds$, where $\lambda = \min_{\alpha \in G} \lambda_\alpha$ (λ_α is the sojourn intensity of the process $\kappa(t)$ at the state $\alpha \in G$). Thus, in the period of time preceding T_0, the processes $\kappa'(t)$ and $\kappa''(t)$ are coupled with a probability not less than $\delta_1 \delta^2 > 0$.

Let us denote by $\upsilon_{\min} = \min_{i,j \in G}(a_i,\ b_j)$ and $T = 2\frac{V_1 - V_0}{\upsilon_{\min}}$.

Jumps are absent after T_0, with a probability $\delta_2 = \lambda \int_{T_0}^{\infty} e^{-\lambda s} ds$. Therefore, for identical $\kappa'(t)$ and $\kappa''(t)$ in a period of time that does not exceed T, the components $\upsilon'(t)$ and $\upsilon''(t)$, where $\upsilon'(t)$ satisfies equation [4.15], derived by the process $\kappa'(t)$, and $\upsilon''(t)$ satisfies equation [4.15], derived by the process $\kappa''(t)$, are coupled in an atom.

Thus, in the period of time preceding $T + T_0$, the processes

$$\overline{\xi}'(t) = (\kappa'(t),\, \upsilon'(t)) \quad \text{and} \quad \overline{\xi}''(t) = (\kappa''(t),\, \upsilon''(t))$$

are coupled with a probability not less than $\delta_2\delta_1\delta^2 > 0$.

II. We now consider the case where $\kappa(t)$ is a semi-Markov process. Consider the independent processes

$$\xi'(t) = (\tau'(t),\, \kappa'(t),\, \upsilon'(t)), \qquad \xi''(t) = (\tau''(t),\, \kappa''(t),\, \upsilon''(t)),$$

where $\kappa'(t)$ and $\kappa''(t)$ are independent semi-Markov processes with embedded chains $\kappa_l{}'$ and $\kappa_l{}''$, respectively, and with the same sojourn intensities as for $\kappa(t)$. We also know that $\upsilon'(t)$ satisfies equation [4.15], governed by the process $\kappa'(t)$, and $\upsilon''(t)$ satisfies equation [4.15], governed by the process $\kappa''(t)$.

By using lemma 4.2, we will show that, under certain conditions, the processes $(\tau'(t),\, \kappa'(t))$ and $(\tau''(t),\, \kappa''(t))$ can be coupled.

Assume that $\mu_{t,x}(ds) = P\{\tau \in ds \,|\, \tau = t, \kappa = x\}$ and that a continuous density function $p_{t,x}(s) > 0$ exists, i.e.

$$\mu_{t,x}(ds) = p_{t,x}(s)\, ds.$$

Also assume that, for certain $T_0 > 0$, we have

$$\inf_{\substack{(t_1,\, x_1),\, (t_2,\, x_2) \\ t_1,\, t_2 \le T_0}} \int_0^{T_0} \min\left(p_{t_1, x_1}(s),\, p_{t_2, x_2}(s)\right) ds = \delta_3 > 0, \qquad [4.29]$$

$$\inf_{t,x} \mu_{t,x}([0,\, T_0]) = \delta_4 > 0.$$

By virtue of [4.29] and lemma 4.2, $(\tau'(t), \kappa'(t))$ and $(\tau''(t), \kappa''(t))$ can be coupled prior to T_0, with a probability not less than $\delta_3\delta_4\delta^2$. Furthermore, by analogy of the Markov case, we wait until the components $\upsilon'(t)$ and $\upsilon''(t)$ are coupled in an atom. If $\max(t_1,\, t_2) > T_0$, then, with a probability not less than

$$\delta_5 = \int_0^{\infty} \left(F_{x_1}(t_1 + T_0 + t) - F_{x_1}(t_1 + t)\right) f_{x_2}(t_2 + t)\, dt$$

we have $\tau'' = 0$, $\tau' \le T_0$, at the epoch of jump of $\kappa''(t)$ and, taking [4.27] into account, we can use the coupling lemma 4.2.

For practical applications, e.g. for the calculation of the stationary reliability of a system, it is necessary to know the function $\overline{\rho}(\alpha, \upsilon)$, that is equal to the sojourn time of a bivariate process $\zeta(t) = (\kappa(t), \upsilon(t))$ at $(\alpha,\, \upsilon)$ (Pogorui 2011b).

This function is determined as follows:

$$\overline{\rho}(\alpha, v) = \int_0^\infty \rho(\tau, \alpha, v)\, d\tau = c\,\rho_\alpha m_\alpha, \quad \alpha \in G,$$

for $x \in X$ we have

$$\overline{\rho}[x, V_1] = \int_0^\infty \rho[\tau, x, V_1]\, d\tau = c\left(\frac{1}{2}\sum_{z\in X}\rho_z b_z m_z^{(2)} + \sum_{z\in X}\rho_z b_z m_z^2 \sum_{k\in X} p_{zk} g_{kx}\right),$$

whereas for $y \in Y$ we have

$$\overline{\rho}[y, V_0] = \int_0^\infty \rho[\tau, y, V_0]\, d\tau = c\left(\frac{1}{2}\sum_{z\in Y}\rho_z b_z m_z^{(2)} + \sum_{z\in Y}\rho_z a_z m_z^2 \sum_{k\in Y} p_{zk} g_{ky}\right),$$

where

$$c^{-1} = \sum_{\beta\in G}\rho_\beta b_\beta m_\beta + \frac{1}{2}\sum_{\beta\in G}\rho_\beta b_\beta m_\beta^{(2)} + \sum_{x\in X}\sum_{z\in X}\rho_z b_z\, m_z^2 \sum_{k\in X} p_{zk} g_{kx}$$

$$+ \sum_{y\in Y}\sum_{z\in Y}\rho_z a_z\, m_z^2 \sum_{k\in Y} p_{zk} g_{ky}.$$

4.2.3. *Stationary distribution of random evolution in semi-Markov media with delaying boundaries*

Now, we will study the problem of section 4.2.2, without considering the balance condition D_4. We will keep the same notation used in section 4.2.2, assuming that conditions D_1–D_3 are satisfied.

Thus, by solving equation [4.17], we have:

$$\rho(\tau, x, v) = f_x(v - b_x\tau)\, e^{-\int_0^\tau r_x(s)ds}, \quad x \in X,$$

$$\rho(\tau, y, v) = f_y(v + a_y\tau)\, e^{-\int_0^\tau r_y(s)ds}, \quad y \in Y.$$

By considering equations [4.18], we will find functions $f_\alpha(\cdot)$, $\alpha \in G$ in the following form:

$$f_\alpha(x) = c_\alpha e^{sx}, \quad c_\alpha > 0, \alpha \in G. \qquad [4.30]$$

By substituting [4.30] into [4.18], we obtain

$$\sum_{\beta\in G} c_\beta \int_0^\infty f_\beta(\tau)\, e^{sk_\beta\tau} d\tau\, p_{\beta\alpha} = c_\alpha, \quad \alpha \in G, \qquad [4.31]$$

where

$$k_\beta = \begin{cases} -a_\beta, & \beta \in X, \\ b_\beta, & \beta \in Y. \end{cases}$$

Denote as $\tilde{f}_\beta\left(sk_\beta\right) = \int_0^\infty f_\beta\left(\tau\right) e^{sk_\beta \tau} d\tau$. Then, it is easily seen that [4.31] has a non-zero solution $c_\alpha, \alpha \in G$ if, and only if, the following condition holds:

$$\begin{vmatrix} -1 & \tilde{f}_{x_1}\left(-a_1 s\right) p_{x_1 x_2} & \cdots & \tilde{f}_{x_1}\left(-a_1 s\right) p_{x_1 y_1} & \cdots & \tilde{f}_{x_1}\left(-a_1 s\right) p_{x_1 y_m} \\ \tilde{f}_{x_2}\left(-a_2 s\right) p_{x_2 x_1} & -1 & \cdots & \tilde{f}_{x_2}\left(-a_2 s\right) p_{x_2 y_1} & \cdots & \tilde{f}_{x_2}\left(-a_2 s\right) p_{x_2 y_m} \\ \tilde{f}_{x_3}\left(-a_3 s\right) p_{x_3 x_1} & \tilde{f}_{x_3}\left(-a_3 s\right) p_{x_3 x_2} & \cdots & \tilde{f}_{x_3}\left(-a_3 s\right) p_{x_3 y_1} & \cdots & \tilde{f}_{x_3}\left(-a_3 s\right) p_{x_3 y_m} \\ \vdots & \vdots & \vdots & \vdots & \ddots & \vdots \\ \tilde{f}_{y_m}\left(b_m s\right) p_{y_m x_1} & \tilde{f}_{y_m}\left(b_m s\right) p_{y_m x_2} & \cdots & \tilde{f}_{y_m}\left(b_m s\right) p_{y_m y_1} & \cdots & -1 \end{vmatrix}$$
$$= 0. \qquad [4.32]$$

It is clear that $s = 0$ is a solution of equation [4.32]. Let $\{s_0 = 0, \, s_1, \ldots, s_l\}$, $l \geq 1$ be a set of solutions of [4.32]. Each s_i, $0 \leq i \leq l$ corresponds to a solution $\{c_\alpha^i, \alpha \in G\}$ of equation [4.31]. For example, $s_0 = 0$ corresponds to $c\rho_\alpha$, where c is a normalizing factor. Thus, $\rho\left(\tau,\alpha,v\right)$ will be found in the following form:

$$\rho\left(\tau,x,v\right) = \sum_{i=0}^{l} c_x^i e^{s_i\left(v + a_x \tau\right)} \left(1 - F_x\left(\tau\right)\right), \quad x \in X,$$

$$\rho\left(\tau,y,v\right) = \sum_{i=0}^{l} c_y^i e^{s_i\left(v - b_y \tau\right)} \left(1 - F_y\left(\tau\right)\right), \quad y \in Y.$$

Then, by solving [4.19] and [4.20], we obtain

$$\rho\left[\tau, \, x, \, V_0\right] = \sum_{i=1}^{l} \frac{c_x^i}{s} \left(1 - F_x\left(\tau\right)\right) e^{s_i V_0} \left(e^{a_x s_i \tau} - 1\right) + c\,\rho_x a_x \tau \, \left(1 - F_x\left(\tau\right)\right)$$
$$+ \rho\left[0, \, x, \, V_0\right] \left(1 - F_x\left(\tau\right)\right), \qquad [4.33]$$
$$\rho\left[\tau, \, y, \, V_1\right] = \sum_{i=1}^{l} \frac{c_x^i}{s} \left(1 - F_y\left(\tau\right)\right) e^{s_i V_1} \left(1 - e^{-b_y s_i \tau}\right) + \sigma \rho_y a_y \tau \left(1 - F_y\left(\tau\right)\right)$$
$$+ \rho\left[0, \, y, \, V_1\right] \left(1 - F_y(\tau)\right).$$

Substituting [4.33] in the first two equations of [4.21], we obtain

$$\rho\left[0, \, z, \, V_0\right] = \sum_{i=1}^{l} \frac{e^{s_i V_0}}{s_i} \sum_{x \in X} \left[c_x^i \tilde{f}\left(s a_x\right) \sum_{k \in X} p_{xk} g_{kz} - c_x^i \sum_{k \in X} p_{xk} g_{kz} \right]$$

$$+\sum_{x\in X}\sigma\rho_x a_x\, m_x\sum_{k\in X}p_{xk}g_{kz},\quad z\in X,$$

$$\rho\,[0,\,z,\,V_0]=\sum_{i=1}^{l}\frac{e^{s_iV_1}}{s_i}\sum_{y\in Y}\left[c_y^i\tilde{f}\,(sa_y)\sum_{k\in Y}p_{yk}g_{kz}-c_y^i\sum_{k\in Y}p_{xk}g_{kz}\right]$$

$$+\sum_{y\in Y}\sigma\rho_y a_y\, m_y\sum_{k\in Y}p_{xk}g_{kz},\quad z\in Y.$$

To find the required values $\{c_\alpha^i,\ \alpha\in G\}$, we substitute these expressions into the last two equations of [4.21]. Hence, we have the following condition:

E. *Among solutions* $\{c_\alpha^i,\ \alpha\in G\}$ *of* [4.31], *corresponding to* s_i $i=0,1,\ldots,l$, *that are non-zero* $\{c_\alpha^{i_r},\ \alpha\in G,\ r=0,1,\ldots,d\}$, *satisfy the following equations:*

$$\sum_{r=1}^{d}\sum_{x\in X}c_x^{i_r}\left(\tilde{f}\,(s_{i_r}a_x)-1\right)\sum_{z\in X}g_{xz}p_{zy}+c\sum_{x\in X}a_xm_x\rho_x\sum_{z\in X}g_{kz}p_{zy}$$

$$=\sum_{r=1}^{d}c_y^{i_r}\left(1-\tilde{f}\,(-s_{i_r}b_y)\right)+c\,b_ym_y\rho_y,\quad\forall\,y\in Y,$$

$$\sum_{r=1}^{d}\sum_{y\in Y}c_y^{i_r}\left(\tilde{f}\,(-s_{i_r}b_y)-1\right)\sum_{z\in Y}g_{yz}p_{zx}+c\sum_{y\in Y}a_ym_y\rho_y\sum_{z\in Y}g_{yz}p_{zx}$$

$$=\sum_{r=1}^{d}c_x^{i_r}\left(1-\tilde{f}\,(s_{i_r}a_x)\right)+c\,b_xm_x\rho_x,\quad\forall\,x\in X.$$

Thus, we have proved the following theorem:

THEOREM 4.3.– If conditions D_2 – D_3 and E are satisfied, then the Markov process $\xi(t)$ has the stationary distribution $\rho\,(\cdot)$ with the continuous part:

$$\rho\,(\tau,x,v)=\sigma_1\sum_{r=0}^{d}c_x^{i_r}e^{s_{i_r}(v+a_x\tau)}(1-F_x(\tau)),$$

$$\rho\,(\tau,y,v)=\sigma_1\sum_{r=0}^{d}c_y^{i_r}e^{s_{i_r}(v-b_y\tau)}(1-F_y(\tau)),$$

where $c_\alpha^0=c\rho_\alpha,\quad x\in X,\ y\in Y$, and the singular part (atoms):

$$\rho\,[0,z,V_0]=\sigma_1\left(\sum_{r=1}^{d}\frac{e^{s_{i_r}V_0}}{s_{i_r}}\sum_{x\in X}\left[c_x^{i_r}\tilde{f}\,(s_{i_r}a_x)\sum_{k\in X}p_{xk}g_{kz}-c_x^{i_r}\sum_{k\in X}p_{xk}g_{kz}\right]\right.$$

$$+\sum_{x\in X} c\rho_x a_x m_x \sum_{k\in X} p_{xk} g_{kz} \Bigg), \qquad z \in X,$$

$$\rho[0, z, V_0] = \sigma_1 \left(\sum_{r=1}^{d} \frac{e^{s_{i_r} V_1}}{s_{i_r}} \sum_{y\in Y} \left[c_y^{i_r} \tilde{f}(s_{i_r} a_y) \sum_{k\in Y} p_{yk} g_{kz} - c_y^{i_r} \sum_{k\in Y} p_{xk} g_{kz} \right] \right.$$

$$\left. + \sum_{y\in Y} c\rho_y a_y m_y \sum_{k\in Y} p_{xk} g_{kz} \right), \qquad z \in Y,$$

where σ_1 is determined from the condition $\int_Z \rho(z)\,dz = 1$.

4.3. Stationary efficiency of a system with two unreliable subsystems in cascade and one buffer: the Markov case

4.3.1. *Introduction*

In this section, we study the stationary efficiency, or product transfer effectiveness, of two subsystems that are interconnected in a cascade (series), and where the first subsystem is also connected to a buffer. The buffer delivers product to the second subsystem in the event of the failure of the first subsystem. Both subsystems deliver product to its destination through reliable lines, but these two subsystems are considered unreliable. We model the dynamics of the overall system by using a Markov evolution environment, and we derive design formulas involving the main parameters. We include numerical analysis for some specific cases as well. This basic scheme can be extended in several manners, and as an extension, we analyze the parallel interconnection of these two basic systems.

The availability of repairable systems consisting of several stages (or subsystems) in a cascade, also called series systems, has been investigated by several researchers since the early 1960s, by using jump Markov stochastic processes, with exponential distribution for faultless and restoration periods, i.e. up and down periods, as seen in Chapters 6–8 of Gnedenko and Ushakov (1995) (see also, Barlow *et al.* (1965)). A typical approach to increase the availability of such a series system is to introduce redundancy in some of the subsystems. However, when a subsystem is expensive, such an approach produces prohibitive costs. In this section, we explore another possibility to increase the availability of the system, and as a result, the effectiveness for the transfer of product, by introducing a reliable buffer.

We should mention that the random evolutions approach has been proposed before, for modeling the data fluid flow in a single buffer (Kulkarni 1997). However, such models are aimed at capturing the dynamics of statistical multiplexers (Anick *et al.* 1982) and leaky-bucket mechanisms for congestion control in data communications

(Elwalid and Mitra 1994); see also, Sericola (1998) and Reibman *et al.* (1989) for further related results. The most important performance measure in such cases is the tail probability for the loss of information. Spectral methods and asymptotic results are used for such a purpose, which are different to our solution method. We remark that in our single buffer scheme, there is no loss of information in the buffer, which can be inferred from conditions 1 and 2 in this section. Our buffer is a spare subsystem aimed at increasing the availability of the whole system. Besides the $C(z)$ driving function given in equation [4.49], typical in the random evolution formulation (Kulkarni 1997), we introduce an additional function $f(z)$ in equation [4.51] to capture the functionality of our scheme. These functions, $f(z)$ and $C(z)$, allow us to model more sophisticated systems, such as the parallel system presented in the second half of this section.

The dynamics of this linear system can be captured by a first-order differential equation having a random process. In probabilistic literature, these models are called random evolutions (Korolyuk and Korolyuk 1999). These systems have been analyzed before by using the random evolutions approach. For instance, the cascade of an arbitrary number of subsystems and reservoirs has been investigated in Pogorui (2003), where some asymptotic results have been obtained. Similarly, a cascade of two subsystems and two buffers has been analyzed in Pogorui and Turbin (2002), where a closed-form expression for the effectiveness has been obtained for the so-called "balance" case. However, the analysis becomes too cumbersome to obtain such an effectiveness, and it is not clear which is the fault tolerance gain of the system, by introducing two buffers. In sections 4.3.2–4.3.4, we consider the single system. In section 4.3.2, we elaborate our mathematical modeling. In section 4.3.3, we present our main mathematical results, including a closed-form expression for K (effectiveness), whereas numerical results for several special cases are shown in section 4.3.4.

4.3.2. *Stationary distribution of Markov stochastic evolutions*

In reliability problems, where we like to find the stationary parameters of a system's effectiveness, we need to calculate the stationary distribution of the random processes that are modeling the given system. In particular, we will deal with this problem in the Markov case in this section.

Consider an evolutionary switched process $v(t)$ that is determined by the equation

$$\frac{dv(t)}{dt} = C(\kappa(t), v(t)), \qquad [4.34]$$

where $\kappa(t)$ is a Markov process with phase space $\mathbb{G} = \mathbb{X} \cup \mathbb{Y}$, $\mathbb{X} = \{x_1, \dots, x_n\}$, $\mathbb{Y} = \{y_1, \dots, y_m\}$.

Let us assume that in $\kappa(t)$ there is a homogeneous Markov chain κ_l, $l \geq 1$ embedded, which has a transition probabilities matrix $P = \{p_{\alpha\beta}, \alpha, \beta \in \mathbb{G}\}$,

$p_{\alpha\beta} = P\{\kappa_{l+1} = \beta \,|\, \kappa_l = \alpha\}$, and the holding, or sojourn time τ_α of process $\kappa(t)$ in a state α has a cumulative distribution function $P\{\tau_\alpha \leq t\} = 1 - e^{-\lambda_\alpha t}$.

It is known (Korolyuk and Swishchuk 1995b; Korolyuk and Korolyuk 1999), that equation [4.34] determines the stochastic random evolution in Markov media.

Let the following conditions be satisfied:

– C1) The phase space of the embedded Markov chain κ_l consists of one ergodic class $\mathbb{G}$.

– C2) Now, let us consider real numbers $V_0 < V_1$, $a_i > 0$, $b_j > 0$, $i = 1, \ldots, n$; $j = 1, \ldots, m$, and suppose that the function from the right part of equation [4.34] satisfies:

i) for $x_i \in \mathbb{X}$, $i = 1, \ldots, n$,

$$C(x_i, v) = \begin{cases} -a_i, & V_0 < v \leq V_1 \\ 0, & v = V_0, \end{cases}$$

ii) for $y_j \in \mathbb{Y}$, $j = 1, \ldots, m$,

$$C(y_j, v) = \begin{cases} b_j, & V_0 \leq v < V_1 \\ 0, & v = V_1. \end{cases}$$

Let us consider the two-component process (or bivariate process) $\xi(t) = (\kappa(t), v(t))$ on the phase space $\mathbb{Z} = \mathbb{G} \times [V_0, V_1]$. Our purpose is to investigate the stationary distribution of the Markov process $\xi(t)$.

The process $\xi(t)$ has an infinitesimal operator A, of the following form (Korolyuk and Swishchuk 1995b), (Korolyuk and Korolyuk 1999):

$$A\varphi(\alpha, v) = C(\alpha, v)\frac{d}{dv}\varphi(\alpha, v) + Q\varphi(\alpha, v),$$

where $Q\varphi(\alpha, v) = \lambda_\alpha[P\varphi(\alpha, v) - \varphi(\alpha, v)]$, and $P\varphi(\alpha, v) = \sum_{\beta \in \mathbb{G}} p_{\alpha\beta}\varphi(\beta, v)$.

If process $\{\xi(t)\}$ has a stationary distribution $\rho(\cdot)$, then for any function $\varphi(\cdot)$ belonging to the domain of definition of the operator A, we have:

$$\int_{\mathbb{Z}} A\varphi(z)\rho(dz) = 0. \qquad [4.35]$$

The analysis of properties of the process $\xi(t)$ leads to the conclusion that in points (x, V_0), $x \in \mathbb{X}$ and (y, V_1), $y \in \mathbb{Y}$ of the phase space $\mathbb{Z}$, there are atoms on v of the stationary distribution $\rho(\cdot)$. Henceforth, we shall denote these atoms by $\rho[x, V_0]$, $\rho[y, V_1]$, and its probability density by $\rho(\alpha, v)$, $\alpha \in \mathbb{Z}$, $V_0 < v < V_1$.

Changing the order of integration in [4.35], we obtain expressions for $A^*\rho = 0$, where A^* is the conjugate (dual or adjoint) operator for operator A. Namely,

$$\begin{aligned}
&\int_Z A\varphi(z)\rho(dz) \\
&= \sum_{\alpha \in G} \int_{V_0}^{V_1} \left(C(\alpha, v)\frac{d}{dv}\varphi(\alpha, v) + \lambda_\alpha(P\varphi(\alpha, v) - \varphi(\alpha, v))\right)\rho(\alpha, v)dv \\
&= \sum_{\alpha \in G} (\varphi(\alpha, V_1)C(\alpha, V_1)\rho(\alpha, V_1^-) - \varphi(\alpha, V_0)C(\alpha, V_0)\rho(\alpha, V_0^+) \qquad [4.36] \\
&- \int_{V_0}^{V_1} \varphi(\alpha, v)C(\alpha, v)\frac{d}{dv}\rho(\alpha, v)dv) + \sum_{\alpha \in G} \rho(\alpha, v)(P\varphi(\alpha, v) - \varphi(\alpha, v)) = 0,
\end{aligned}$$

where $\rho(\alpha, V_0^+) = \lim_{v \downarrow V_0} \rho(\alpha, v)$ and $\rho(\alpha, V_1^-) = \lim_{v \uparrow V_1} \rho(\alpha, v)$.

As a result, for the density we have

$$C(\alpha, v)\frac{d}{dv}\rho(\alpha, v) + Q^*\rho(\alpha, v) = 0, \quad \alpha \in G, V_0 < v < V_1, \qquad [4.37]$$

where Q^* is the conjugate operator of Q:

$$Q^*\rho(\alpha, v) = \sum_{\beta \in G} \lambda_\beta p_{\beta\alpha}\rho(\beta, v) + \lambda_\alpha \rho(\alpha, v).$$

Denote by

$$\boldsymbol{\rho}[V_0] = (\rho[x_1, V_0], \ldots, \rho[x_n, V_0], 0, \ldots, 0),$$

$$\boldsymbol{\rho}[V_1] = (0, \ldots, 0, \rho[y_1, V_1], \ldots, \rho[y_m, V_1])$$

$(n + m)$-dimensional vectors, and by

$$C = \mathrm{diag}(-a_1, -a_2, \ldots, -a_n, b_1, b_2, \ldots, b_m)$$

the $(n+m) \times (n+m)$ diagonal matrix. We shall denote by Q, the matrix representation of operator Q. Thus,

$$Q = \begin{pmatrix} -\lambda_{x_1} & \lambda_{x_1}p_{x_1x_2} & \cdots & \lambda_{x_1}p_{x_1y_m} \\ \lambda_{x_2}p_{x_2x_1} & -\lambda_{x_2} & \cdots & \lambda_{x_2}p_{x_2y_m} \\ \vdots & \vdots & \ddots & \vdots \\ \lambda_{y_m}p_{y_mx_1} & \lambda_{y_m}p_{y_mx_2} & \cdots & -\lambda_{y_m} \end{pmatrix},$$

where it is assumed that $p_{\alpha\alpha} = 0$, for all $\alpha \in G$.

For the atoms, it follows from [4.36] that

$$\boldsymbol{\rho}[V_0]Q = \boldsymbol{\rho}(V_0^+)C, \qquad \boldsymbol{\rho}[V_1]Q = \boldsymbol{\rho}(V_1^-)C. \qquad [4.38]$$

We write equations [4.37] in the vector–matrix form:

$$\frac{d}{dv}\boldsymbol{\rho}(v)C + \boldsymbol{\rho}(v)Q = \mathbf{0}, \quad v \in (V_0, V_1) \qquad [4.39]$$

Now, by solving [4.39] we obtain

$$\boldsymbol{\rho}(v) = \boldsymbol{\rho}(V_0^+)\exp(QC^{-1}(V_0 - v)), \quad v \in (V_0, V_1). \qquad [4.40]$$

The measure $\boldsymbol{\rho}(V_0^+)$ and its atoms $\boldsymbol{\rho}[V_0]$ and $\boldsymbol{\rho}[V_1]$ can be obtained from equations [4.38]. These values are determined up to a normalizing factor c. This fact is a consequence of condition C1, implying that the equation $Q\mathbf{x} = \mathbf{0}$ has a unique solution $\mathbf{x} = \mathbf{1}$ (up to a normalizing factor), where $\mathbf{1} = (1, \ldots, 1)^*$ is a $(n+m)$-vector.

Hence, matrix formulas [4.38] contain $2(n+m) - 2$ independent equations and vectors $\boldsymbol{\rho}(V_0^+)$, $\boldsymbol{\rho}[V_0]$, $\boldsymbol{\rho}[V_1]$ contain $2(n+m)-1$ independent variables. This is since the vector $\boldsymbol{\rho}(V_1^-)$, from formula [4.40], is uniquely determined in terms of vector $\boldsymbol{\rho}(V_0^+)$.

The normalizing factor c can be obtained from the following equation:

$$\int_{V_0}^{V_1} \boldsymbol{\rho}(v)\mathbf{1}dv + (\boldsymbol{\rho}[V_0] + \boldsymbol{\rho}[V_1])\mathbf{1} = 1. \qquad [4.41]$$

Thus, the following theorem has been proved.

THEOREM 4.4.– If conditions C1 and C2 hold, then the Markov process $\xi(t) = (\kappa(t), v(t))$ has a stationary distribution $\rho(\cdot)$ with a density determined by the formula $\boldsymbol{\rho}(v) = \boldsymbol{\rho}(V_0^+)\exp(QC^{-1}(V_0 - v)), v \in (V_0, V_1)$, and atoms $\boldsymbol{\rho}[V_0], \boldsymbol{\rho}[V_1]$ determined by equation [4.38].

EXAMPLE 4.3.– Let us consider the special case when $n = 1$, i.e. $X = \{x\}$, $Y = \{y_1, y_2, \ldots, y_m\}$. Then,

$$\boldsymbol{\rho}[V_0] = (\rho[x, V_0], 0, \ldots, 0),$$

$$\boldsymbol{\rho}[V_1] = (0, \rho[y_1, V_1], \ldots, \rho[y_m, V_1])$$

and matrices Q, C are given by

$$Q = \begin{pmatrix} -\lambda_x & \lambda_x p_{xy_1} & \dots & \lambda_x p_{xy_m} \\ \lambda_{y_1} p_{y_1 x} & -\lambda_{y_1} & \dots & \lambda_{y_1} p_{y_1 y_m} \\ \vdots & \vdots & \ddots & \vdots \\ \lambda_{y_m} p_{y_m x} & \lambda_{y_m} p_{y_m y_1} & \dots & -\lambda_{y_m} \end{pmatrix}, \quad C = \begin{pmatrix} -a & 0 & \dots & 0 \\ 0 & b_1 & \dots & 0 \\ \vdots & \vdots & \ddots & \vdots \\ 0 & 0 & \dots & b_m \end{pmatrix}.$$

The first formula of [4.38] implies

$$\rho(x, V_0^+) = \frac{\lambda_x \rho[x, V_0]}{a}, \quad \rho(y_1, V_0^+) = \frac{\lambda_x p_{xy_1} \rho[x, V_0]}{b_1},$$

$$\dots, \rho(y_m, V_0^+) = \frac{\lambda_x p_{xy_m} \rho[x, V_0]}{b_m}.$$

The second equation of [4.38] is of the following form:

$$(0, \rho[y_1, V_1], \dots, \rho[y_m, V_1])Q = \boldsymbol{\rho}(V_0^+) \exp(QC^{-1}(V_0 - v))C. \qquad [4.42]$$

Denote by Q_Y and C_Y matrices, which are obtained from matrices Q and C after the removal of the first column and the first line, and assume $\boldsymbol{\rho}_Y(V_1^-) = (\rho(y_1, V_1^-), \dots, \rho(y_m, V_1^-))$.

The condition C2 implies the existence of an inverse matrix Q_Y^{-1} corresponding to matrix Q_Y. Consequently, taking into account [4.38] we have

$$(\rho[y_1, V_1], \dots, \rho[y_m, V_1]) = (\rho(y_1, V_1^+), \dots, \rho(y_1, V_m^+))C_Y Q_Y^{-1}. \qquad [4.43]$$

Now, since $\boldsymbol{\rho}_Y(V_1^-)$ is uniquely expressed in terms of $\boldsymbol{\rho}_Y(V_0^+)$, we obtain the stationary distribution of $\xi(t)$ with density

$$\boldsymbol{\rho}(v) = \left(\frac{\lambda_x c}{a}, \frac{\lambda_{y_1} c}{b_1}, \dots, \frac{\lambda_{y_1} c}{b_1}\right) \exp(QC^{-1}(V_0 - v)), \; v \in (V_0, V_1). \quad [4.44]$$

The atoms are determined by formula [4.44], and the normalizing factor $c = \rho[x, V_0]$ is determined by [4.41].

4.3.3. *Stationary efficiency of a system with two unreliable subsystems in cascade and one buffer*

In this section, as an important example of the previous section, we analyze a series system with two subsystems and a single buffer, as shown in Figure 4.1. We also obtain a closed-form expression for the effectiveness of this system and have established the fault tolerance gain provided by the introduction of the buffer.

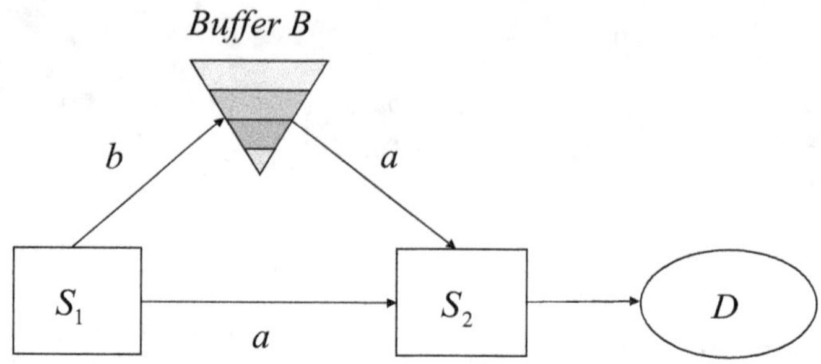

Figure 4.1. *A system of two unreliable subsystems, say S_1 and S_2, connected in series with a buffer B with capacity V. For a color version of this figure, see www.iste.co.uk/pogorui/random1.zip*

Let us describe the main elements of this system. The two subsystems, or transmission subsystems, transferring information are denoted as S_1, S_2, with a reservoir (storage or buffer) B, which can accumulate a maximum of V amount of information or product (its capacity).

All of the interconnection lines are considered reliable, and the time for failure detection is negligible. The non-failure operating time interval $\tau_i^{(1)}$, of subsystem S_i, is an exponential distributed random variable with a failure rate or parameter $\lambda_i > 0$, $i = 1, 2$, (up period), and the renewal (downtime or restoration period) time $\tau_i^{(2)}$ of subsystem S_i is also an exponential distributed random variable with a repair rate or parameter μ_i, $i = 1, 2$.

The system functionality is as follows:

1) If subsystems S_1 and S_2 are working properly and the reservoir B is full, then the product is transferred from subsystem S_1 to subsystem S_2 and then to its destination D (or consumer) without any problem, with velocity a.

2) If subsystems S_1, S_2 are working and the buffer B is not full, then the product goes from subsystem S_1 to subsystem S_2, and then to its destination D, with rapidity a. In addition, the product goes from subsystem S_1 to the buffer B, with velocity b.

3) If subsystem S_1 is not working, subsystem S_2 is working and the buffer B is not empty, then the product flows from the buffer B to subsystem S_2, and then to its destination D, with velocity a.

4) If subsystem S_2 is not working, then the product is not transferred to its destination D. However, if the subsystem S_1 is working and the buffer B is not full, then the product flows from S_1 to B, with velocity b.

5) If subsystem S_1 is not working, subsystem S_2 is working and the buffer B is empty, then the product does not flow to the destination D.

Typically, the provisioning of fault tolerance has been done by adding redundant spares to the system. However, in this scheme we are providing some degree of fault tolerance by introducing a buffer. As we know, data memories are becoming cheaper on the market and at present, they are very reliable because of the technology itself and by the addition of error correcting codes in their electronic design.

Now, let us denote as $V(T)$, the volume or amount of product delivered to destination D in the time interval [0,T]. Thus, we can define $K = \lim_{T\to\infty} \frac{V(T)}{T}$ (when it exists) as the steady-state parameter for system performance (or effectiveness). Our main purpose in this subsection is to determine K as a function of the system parameters a, b, V, λ_1, λ_2, μ_1 and μ_2.

4.3.4. *Mathematical model*

Let us introduce the following stochastic process $\{\chi(t)\}$, such that

$$\chi(t) := \begin{cases} 00, & \text{if } S_1 \text{ and } S_2 \text{ are not working;} \\ 01, & \text{if } S_2 \text{ is working, and } S_1 \text{ is not working;} \\ 10, & \text{if } S_1 \text{ is working, and } S_2 \text{ is not working;} \\ 11, & \text{if } S_1 \text{ and } S_2 \text{ are working.} \end{cases} \qquad [4.45]$$

The stochastic process $\chi(t)$ is a Markov process on the phase space (or states) $\mathcal{Z} = \{00, 01, 10, 11\}$. To simplify notation, we make the following correspondence $00 \Leftrightarrow 0$, $01 \Leftrightarrow 1$, $10 \Leftrightarrow 2$, and $11 \Leftrightarrow 3$. Hence, the generating operator (or matrix) of $\chi(t)$ can be written as (Korolyuk and Korolyuk 1999):

$$Q = q\,[P - I] = \begin{bmatrix} -(\mu_1+\mu_2) & \mu_2 & \mu_1 & 0 \\ \lambda_2 & -(\mu_1+\lambda_2) & 0 & \mu_1 \\ \lambda_1 & 0 & -(\lambda_1+\mu_2) & \mu_2 \\ 0 & \lambda_1 & \lambda_2 & -(\lambda_1+\lambda_2) \end{bmatrix}, \qquad [4.46]$$

where $q = [q_i\delta_{ij};\ i,j \in \{0,1,2,3\}]$ is a diagonal matrix of sojourn times intensities of different states, and $q_0 = \mu_1 + \mu_2$, $q_1 = \mu_1 + \lambda_2$, $q_2 = \lambda_1 + \mu_2$, $q_3 = \lambda_1 + \lambda_2$. Here, as usual, the Kronecker's delta is defined as

$$\delta_{ij} := \begin{cases} 1, & i = j; \\ 0, & i \neq j. \end{cases} \qquad [4.47]$$

The elements of matrix P are the transition probabilities of the Markov chain embedded in the Markov process $\chi(t)$, i.e.

$$P = \begin{bmatrix} 0 & \frac{\mu_2}{\mu_1+\mu_2} & \frac{\mu_1}{\mu_1+\mu_2} & 0 \\ \frac{\lambda_2}{\mu_1+\lambda_2} & 0 & 0 & \frac{\mu_1}{\mu_1+\lambda_2} \\ \frac{\lambda_1}{\lambda_1+\mu_2} & 0 & 0 & \frac{\mu_2}{\lambda_1+\mu_2} \\ 0 & \frac{\lambda_1}{\lambda_1+\lambda_2} & \frac{\lambda_2}{\lambda_1+\lambda_2} & 0 \end{bmatrix}. \qquad [4.48]$$

Consider a function $C(z)$ on the space $\mathbb{Z} = \{00, 01, 10, 11\} \times [0, V]$, which is defined as

$$C(z) := \begin{cases} b, & z = \{11, v\},\ z = \{10, v\},\ v < V; \\ -a, & z = \{01, v\},\ v > 0; \\ 0, & \text{in other cases.} \end{cases} \qquad [4.49]$$

Let $\nu(t)$ be the amount of information (or product, e.g., oil) in the reservoir B at time t, hence it is easily verified that $\nu(t)$ obeys the ordinary differential equation

$$\frac{d\,\nu(t)}{dt} = C(\chi(t), \nu(t)), \qquad [4.50]$$

with initial conditions $\nu(0) = \nu_0 \in [0, V]$. It can be said that equation [4.50] determines the random evolution of the system in the Markov medium $\chi(t)$ (Korolyuk and Korolyuk 1999).

Now, let us define the function $f(z)$ on $\mathbb{Z}$ as follows:

$$f(z) := \begin{cases} a, & \text{if } z = \{11, v\},\ 0 \le v \le V; \\ a, & \text{if } z = \{01, v\},\ v > 0; \\ 0, & \text{in other cases.} \end{cases}, \qquad [4.51]$$

Assume now that the joint stochastic processes with a two-dimensional phase space $\xi(t) = (\chi(t), \nu(t))$. Then, we can state the following equality:

$$K = \lim_{T\to\infty} \frac{V(T)}{T} = \lim_{T\to\infty} \frac{1}{T} \int_0^T f(\xi(t))\, dt. \qquad [4.52]$$

It follows from ergodic theory (Gikhman and Skorokhod 1975; Korolyuk and Turbin 1993; Reed and Simon 1972) that if the process $\xi_i(t)$ has a stationary distribution, say $\rho(\cdot)$, then

$$\lim_{T\to\infty}\frac{1}{T}\int_0^T f(\xi(t))\,dt = \int_{\mathbb{Z}} f(z)\,d\rho(z). \qquad [4.53]$$

Hence, by using equation [4.52] we have

$$K = \int_{\mathbb{Z}} f(z)\,d\rho(z) = \int_{\mathbb{Z}} f(z)\,\rho(z)\,dz. \qquad [4.54]$$

In summary, the calculation of parameter K can be reduced to the calculation of the stationary distribution ρ, of the process $\xi(t)$.

4.3.5. *Main mathematical results*

The sojourn time distribution functions, say $Q_{ij}(t)$, have the following form for the different states: $Q_{00}(t) = 1 - e^{-(\mu_1+\mu_2)t}$, $Q_{01}(t) = 1 - e^{-(\mu_1+\lambda_2)t}$, $Q_{10}(t) = 1 - e^{-(\lambda_1+\mu_2)t}$ and $Q_{11}(t) = 1 - e^{-(\lambda_1+\lambda_2)t}$. Now, denote by $q_x(t) = \frac{dQ_x(t)}{dt}$ and $r_x = \frac{q_x(t)}{1-Q_x(t)}$ for all $x \in \mathcal{Z}$, i.e. $r_{00} = \mu_1 + \mu_2$, $r_{01} = \mu_1 + \lambda_2$, $r_{10} = \lambda_1 + \mu_2$ and $r_{11} = \lambda_1 + \lambda_2$. Then, the two-component process $\xi(t) = (\nu(t), \chi(t))$ is a Markov process with the generator (Korolyuk and Korolyuk 1999; Pogorui and Turbin 2002)

$$A\,\varphi(x,v) = C(x,v)\frac{\partial}{\partial v}\varphi(x,v) + r_x[P\,\varphi(x,v) - \varphi(x,v)], \qquad [4.55]$$

where $P\varphi(x,v) = \sum_{y\in\mathcal{Z}} p_{xy}\varphi(y,v)$, or equivalently

$$A\,\varphi(x,v) = C(x,v)\frac{\partial}{\partial v}\varphi(x,v) + Q\,\varphi(x,v)]. \qquad [4.56]$$

Hence, since $\rho(\cdot)$ is the stationary distribution of the process $\xi(t)$ for every function $\varphi(\cdot)$ belonging to the domain of the operator A, we have

$$\int_{\mathbb{Z}} A\,\varphi(z)\rho(dz) = 0. \qquad [4.57]$$

CONDITION 4.1.– Since at atoms $(01,0)$, $(10,V)$ the subsystems S_1 and S_2 are not involved in the functionality of the system, we assume that they are reliable, i.e. $\lambda_1 = \lambda_2 = 0$ at $(01,0)$, $(10,V)$.

The analysis of the properties of the process $\xi(t)$ leads to the conclusion that the stationary distribution ρ has atoms at the points $(01,0)$, $(10,V)$ and $(11,V)$, and we denote them by $\rho[01,V]$, $\rho[10,V]$ and $\rho[11,V]$, respectively. The continuous part of the stationary distribution ρ, is denoted as $\rho(ij,v)$; $i,j \in \{0,1\}$.

Let us write equation [4.57] in more detail, as follows:

$$\int_{\mathbb{Z}} A\varphi(z)\rho(dz) = \int_0^V \Big\{ \left[-(\mu_1+\mu_2)\varphi(00,v) + \mu_2\varphi(01,v) + \mu_1\varphi(10,v)\right]\rho(00,v)$$
$$+\left[C(01,v)\frac{\partial\varphi(01,v)}{\partial v} + \lambda_2\varphi(00,v) - (\lambda_2+\mu_1)\varphi(01,v) + \mu_1\varphi(11,v)\right]\rho(01,v)$$
$$+\left[C(10,v)\frac{\partial\varphi(10,v)}{\partial v} + \lambda_1\varphi(00,v) - (\lambda_1+\mu_2)\varphi(10,v) + \mu_2\varphi(11,v)\right]\rho(10,v)$$
$$+\left[C(11,v)\frac{\partial\varphi(11,v)}{\partial v} + \lambda_1\varphi(01,v) - (\lambda_1+\lambda_2)\varphi(11,v) + \lambda_2\varphi(10,v)\right]\rho(11,v)\Big\}dv$$
$$+(-\varphi(01,0)+\varphi(11,0))\,\mu_1\rho[01,0] + (-\varphi(10,V)+\varphi(11,V))\,\mu_2\rho[10,V]$$
$$+(\lambda_1\varphi(01,V) + \lambda_2\varphi(10,V) - (\lambda_1+\lambda_2)\varphi(11,V))\,\rho[11,V] = 0. \qquad [4.58]$$

Now, let A^* be the conjugate operator of A. Then, by changing the order of integration in equation [4.58], we can obtain the following expressions for the continuous part of $A^*\rho = 0$:

$$\begin{cases} a\dfrac{\partial}{\partial v}\rho(01,v) + \mu_2\rho(00,v) + \lambda_1\rho(11,v) - (\lambda_2+\mu_1)\rho(01,v) = 0 \\ -b\dfrac{\partial}{\partial v}\rho(10,v) + \mu_1\rho(00,v) + \lambda_2\rho(11,v) - (\lambda_1+\mu_2)\rho(10,v) = 0 \\ -b\dfrac{\partial}{\partial v}\rho(11,v) + \mu_1\rho(01,v) + \mu_2\rho(10,v) - (\lambda_1+\lambda_2)\rho(11,v) = 0 \\ -(\mu_1+\mu_2)\rho(00,v) + \lambda_2\rho(01,v) + \lambda_1\rho(10,v) = 0. \end{cases} \qquad [4.59]$$

Hence, it follows that

$$a\rho(01,v) - b\rho(10,v) - b\rho(11,v) = c_0 = constant. \qquad [4.60]$$

Similarly, from equation [4.57], we obtain

$$\begin{cases} a\rho(01,0^+) - \mu_1\rho[01,0] = 0 \\ \mu_1\rho[01,0] - b\rho(11,0^+) = 0 \\ \rho(10,0^+) = 0 \end{cases}, \qquad [4.61]$$

and

$$\begin{cases} -a\rho(01,V^-) + \lambda_1\rho[11,V] &= 0 \\ b\rho(11,V^-) + \mu_2\rho[10,V] - (\lambda_1+\lambda_2)\rho[11,V] &= 0 \\ b\rho(10,V^-) - \mu_2\rho[10,V] + \lambda_2\rho[11,V] &= 0 \end{cases}. \qquad [4.62]$$

In these equations, we have defined the notation

$$\rho(ij,0^+) := \lim_{v\downarrow 0}\rho(ij,v) \quad \text{and} \quad \rho(ij,V^-) := \lim_{v\uparrow V}\rho(ij,v)$$

for all $i,j \in \{0,1\}$.

Combining equations [4.61] and [4.62], we obtain

$$\begin{cases} a\rho(01,0^+) - b\rho(10,0^+) - b\rho(11,0^+) = 0 \\ a\rho(01,V^-) - b\rho(10,V^-) - b\rho(11,V^-) = 0. \end{cases} \qquad [4.63]$$

It follows from equations [4.61] and [4.63] that $c_0 = 0$ in equation [4.60], so

$$a\rho(01,v) - b\rho(10,v) - b\rho(11,v) = 0. \qquad [4.64]$$

Hence, the first two equations of the set of equations [4.59] can be written in the following form:

$$\begin{cases} \dfrac{\partial\rho(01,v)}{\partial v} = c_{11}\rho(01,v) + c_{12}\rho(10,v) \\ \dfrac{\partial\rho(10,v)}{\partial v} = c_{21}\rho(01,v) + c_{22}\rho(10,v), \end{cases} \qquad [4.65]$$

where

$$\begin{aligned} c_{11} &= -\frac{\lambda_1}{b} + \frac{\mu_1\lambda_2}{a(\mu_1+\mu_2)} + \frac{\mu_1}{a}, \\ c_{12} &= \frac{\mu_1\lambda_1}{a(\mu_1+\mu_2)}, \\ c_{21} &= \frac{a\lambda_2}{b^2} + \frac{\mu_1\lambda_2}{b(\mu_1+\mu_2)}, \\ c_{22} &= -\frac{\lambda_2}{b} - \frac{\mu_2\lambda_1}{b(\mu_1+\mu_2)} - \frac{\mu_2}{b}. \end{aligned} \qquad [4.66]$$

Solving the set of equations [4.65], we obtain

$$\begin{cases} \rho(01,v) &= c_1\, e^{\delta_1 v} + c_2\, e^{\delta_2 v} \\ \rho(10,v) &= c_1 \dfrac{1}{c_{12}}(\delta_1 - c_{11})\, e^{\delta_1 v} + c_2 \dfrac{1}{c_{12}}(\delta_2 - c_{11})\, e^{\delta_2 v}, \end{cases} \qquad [4.67]$$

where

$$\delta_1 = \frac{c_{11} + c_{22} + \sqrt{(c_{22} - c_{11})^2 + 4c_{12}c_{21}}}{2} \qquad [4.68]$$

$$\delta_2 = \frac{c_{11} + c_{22} - \sqrt{(c_{22} - c_{11})^2 + 4c_{12}c_{21}}}{2}. \qquad [4.69]$$

We know from equation [4.61] that $\rho(10, 0^+) = 0$, and from the second equation in the set of equations [4.67], we have

$$c_2 = c_1\, \frac{c_{11} - \delta_1}{\delta_2 - c_{11}}. \qquad [4.70]$$

Hence, the set of equations [4.67] can be rewritten as

$$\begin{cases} \rho(01,v) &= c \left(e^{\delta_1 v} + \dfrac{c_{11} - \delta_1}{\delta_2 - c_{11}}\, e^{\delta_2 v} \right) \\ \rho(10,v) &= \dfrac{c}{c_{12}}(\delta_1 - c_{11}) \left(e^{\delta_1 v} - e^{\delta_2 v}\right). \end{cases} \qquad [4.71]$$

The factor c can be calculated from the normalization equation

$$\int_{\mathbb{Z}} \rho(z)\, dz = 1. \qquad [4.72]$$

For such a purpose, we need to find a few more results for the stationary distribution $\rho(\cdot)$. For instance, by rewriting equation [4.64] and using equation [4.71], we have the solution for $\rho(11, v)$:

$$\rho(11, v) = \frac{a}{b} \rho(01, v) - \rho(10, v). \qquad [4.73]$$

Similarly, by rewriting the last equation of the set of equations [4.59], we obtain the solution for $\rho(00, v)$ as:

$$\rho(00, v) = \frac{\lambda_2}{\mu_1 + \mu_2} \rho(01, v) + \frac{\lambda_1}{\mu_1 + \mu_2} \rho(10, v). \qquad [4.74]$$

The corresponding equations for atoms of the stationary distribution ρ of $\xi(t)$ can be obtained from equations [4.61] and [4.62]. For instance, $\rho[01,0]$ can be obtained directly from the first equation of the set of equations [4.61], and it is given by

$$\rho[01,0] = \frac{a}{\mu_1}\rho(01,0^+). \qquad [4.75]$$

In the same manner,

$$\rho[11,V] = \frac{a}{\lambda_1}\rho(01,V^-), \qquad [4.76]$$

$$\rho[10,V] = \frac{b}{\mu_2}\rho(10,V^-) + \frac{a\lambda_2}{\lambda_1\mu_2}\rho(01,V^-), \qquad [4.77]$$

are obtained from the set of equations [4.62].

Therefore, we give the following closed-form expression for c:

$$\begin{aligned}\frac{1}{c} &= \left(\frac{e^{\delta_1 V}-1}{\delta_1}\right)\left(1+\frac{a}{b}+\frac{\lambda_2}{\mu_1+\mu_2}+\frac{\delta_1-c_{11}}{c_{12}}\frac{\lambda_1}{\mu_1+\mu_2}\right) \\ &+ \left(\frac{e^{\delta_2 V}-1}{\delta_2}\right)\left(\frac{c_{11}-\delta_1}{\delta_2-c_{11}}\left(1+\frac{a}{b}+\frac{\lambda_2}{\mu_1+\mu_2}\right)-\frac{\delta_1-c_{11}}{c_{12}}\frac{\lambda_1}{\mu_1+\mu_2}\right) \\ &+ a\frac{c_{11}-\delta_1}{\delta_2-c_{11}}\left(\frac{1}{\mu_1}+\left(\frac{1}{\lambda_1}+\frac{\lambda_2}{\mu_2\lambda_1}\right)e^{\delta_2 V}\right)+\frac{a}{\mu_1} \\ &+ \frac{a}{\lambda_1}\left(\frac{\lambda_2}{\mu_2}+1\right)e^{\delta_1 V}+\frac{b}{\mu_2}\frac{\delta_1-c_{11}}{c_{12}}\left(e^{\delta_1 V}-e^{\delta_2 V}\right).\end{aligned} \qquad [4.78]$$

Now, from the formula

$$K = a\int_0^V (\rho(11,v)+\rho(01,v))\,dv + a\rho[11,V], \qquad [4.79]$$

we have

$$\begin{aligned}\frac{K}{a\,c} &= \frac{e^{\delta_1 V}-1}{\delta_1}\left(1+\frac{a}{b}-\frac{\delta_1-c_{11}}{c_{12}}\right) \\ &+ \frac{e^{\delta_2 V}-1}{\delta_2}\left(\frac{c_{11}-\delta_1}{\delta_2-c_{11}}\left(1+\frac{a}{b}\right)+\frac{\delta_1-c_{11}}{c_{12}}\right) \\ &+ \frac{a}{\lambda_1}\left(e^{\delta_1 V}+\frac{c_{11}-\delta_1}{\delta_2-c_{11}}e^{\delta_2 V}\right).\end{aligned} \qquad [4.80]$$

4.3.6. Numerical results for the symmetric case

It is interesting to mention that when $V = 0$, we can obtain an efficiency K, that we call K_0, as

$$K_0 = \frac{a}{\left(1 + \frac{\lambda_1}{\mu_1} + \frac{\lambda_2}{\mu_2}\right)}. \qquad [4.81]$$

On the other hand, it is well known that the availability of a series system, say A_v, similar to the system in Figure 4.1, is given by

$$A_v = \left(\frac{1/\lambda_1}{1/\lambda_1 + 1/\mu_1}\right)\left(\frac{1/\lambda_2}{1/\lambda_2 + 1/\mu_2}\right) = \frac{1}{\left(1 + \frac{\lambda_1}{\mu_1} + \frac{\lambda_2}{\mu_2} + \frac{\lambda_1\lambda_2}{\mu_1\mu_2}\right)}. \qquad [4.82]$$

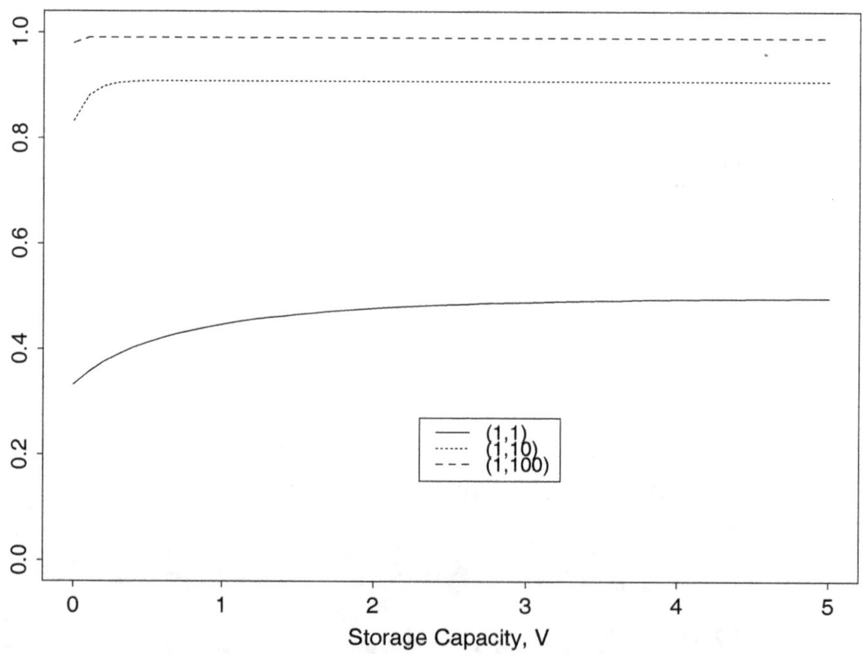

Figure 4.2. *Efficiency parameter K as a function of reservoir size V for different (λ, μ) combinations. We have fixed $a = 1$ and $b = 1$*

Hence, the efficiency of this series system, say K_0^*, is given by aA_v. We should remark that $K_0^* \cong K_0$ for most practical systems since $\lambda_i \ll \mu_i$. However, they are slightly different due to condition 4.1.

In order to gain some insights for the behavior of this system, we analyze the numerical results for the symmetric case, i.e. $\lambda_1 = \lambda_2 = \lambda$ and $\mu_1 = \mu_2 = \mu$.

In Figure 4.2, we plot the efficiency K as a function of reservoir size V, for three different values of the ratio λ/μ, namely $\lambda = 1$, and $\mu = 100$, 10 and 1. We assume $a = b = 1$. We can observe that K increases as the buffer capacity V increases. The maximum value for K is achieved for smaller buffer capacities, when the ratio λ/μ is smaller (systems with higher availabilities). However, the fault tolerance gain provided by the buffer ($K_{max} - K_0^*$) is also smaller in these cases.

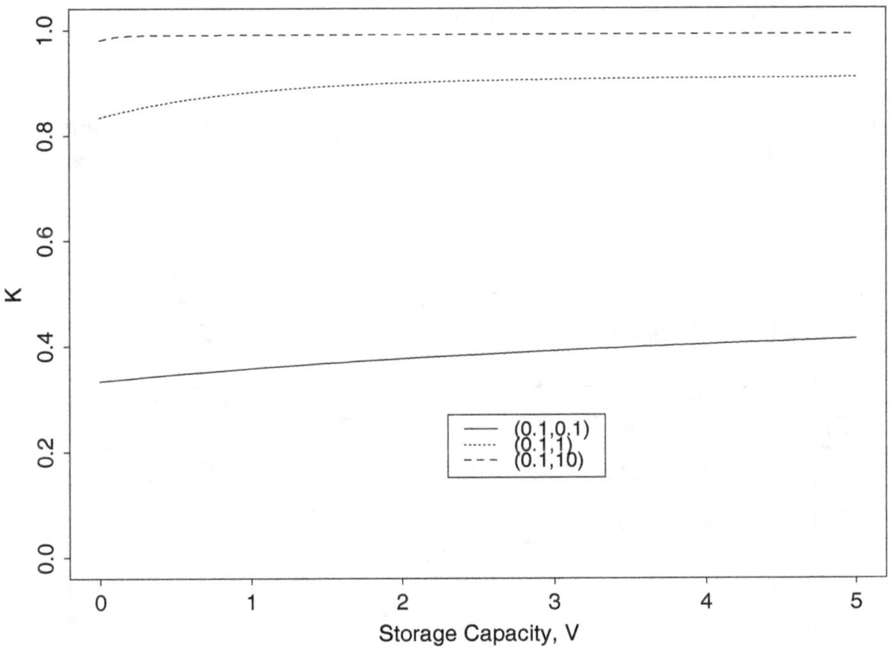

Figure 4.3. *Efficiency parameter K as a function of reservoir size V for different (λ, μ) combinations. We have fixed $a = 1$ and $b = 1$*

A similar behavior can be observed in Figure 4.3, where we plot K for three different values of the ratio λ/μ, namely $\lambda = 0.1$, and $\mu = 0.1$, 1, and 10. The influence of the parameter λ on K is that the maximum is achieved at larger buffer capacities.

This is stated with more detail in Table 4.1, where we can observe that the maximum value of K is not the same at $V = 5$, being smaller when $\lambda = 0.1$ for the same λ/μ rates. This means that larger buffer capacities are required to achieve the same fault tolerance gain, when the system has smaller values of λ.

λ	μ	λ/μ	max K at $V = 5$	min $K = K_0$	K_0^*
1	100	0.01	0.990099	0.9803922	0.980296
1	10	0.1	0.909090	0.8333333	0.8264463
1	1	1	0.49775	0.3333333	0.25
0.1	10	0.01	0.990099	0.9803922	0.980296
0.1	1	0.1	0.9085053	0.8333333	0.8264463
0.1	0.1	1	0.4132633	0.3333333	0.25

Table 4.1. *Maximum and minimum values of K for different λ/μ ratios. We have fixed $a = 1$ and $b = 1$*

In Table 4.2, we state the effect on the maximum K, by varying the parameter b. It seems that the variation around 1 of this parameter is not important for λ/μ ratios of 0.1 and 0.01. It appears important for cases of low availability, and in such cases $a \cong b$ is recommended.

b	λ	max K when $\mu = 1$	max K when $\mu = 10$	max K when $\mu = 100$
0.1	1	0.3413731	0.9090294	0.990099
0.3	1	0.4196899	0.9090909	0.990099
0.6	1	0.4876704	0.9090909	0.990099
0.9	1	0.496913	0.9090909	0.990099
1.1	1	0.4982768	0.9090909	0.990099
2.0	1	0.4995211	0.9090909	0.990099
5.0	1	0.4998276	0.9090909	0.990099
10	1	0.4998797	0.9090909	0.990099
100	1	0.4999138	0.9090909	0.990099
b	λ	max K when $\mu = 0.1$	max K when $\mu = 1$	max K when $\mu = 10$
0.1	0.1	0.3414187	0.8989422	0.990099
0.3	0.1	0.3859017	0.9076991	0.990099
0.6	0.1	0.4050227	0.9083301	0.990099
0.9	0.1	0.4118787	0.9084797	0.990099
1.1	0.1	0.4143976	0.9085254	0.990099
2.0	0.1	0.4194981	0.9086085	0.990099
5.0	0.1	0.4232016	0.9086621	0.990099
10	0.1	0.4244218	0.9086788	0.990099
100	0.1	0.4255116	0.9086933	0.990099

Table 4.2. *Maximum values of K at $V = 5$ for different λ/μ ratios and different values of b. We have fixed $a = 1$*

4.4. Application of random evolutions with delaying barriers to modeling control of supply systems with feedback: the semi-Markov switching process

We will calculate the stationary efficiency of a supply system with feedback. The system considered consists of one or more unreliable flows, or flows with interruptions (or aggregates) and a corresponding number of reservoirs (or buffers). As we will see, these systems can be modeled by random evolutions in Markov or semi-Markov environments, and the calculation of the stationary efficiency of such systems can often be reduced to finding the stationary distribution of the corresponding random evolution.

Some of the first investigations on the reliability theory of systems with storages or reservoirs were done by Sevastyanov (1962), Buzacott (1967) and Cherkesov (1974). These works used, the technique of jump Markov processes. More works related to this topic are given in Kulkarni (1997), Malathornas *et al.* (1983) and Mitra (1998), among others. In some works, for instance Korolyuk and Turbin (1982), the requirement of exponential distributions of faultless operation and restoration is removed and semi-Markov models of operation systems with reservoirs are studied.

A limitation of the methods mentioned above is the assumption of a condition stating that the reserves are added instantaneously in the reservoirs. This is a rather strong assumption. In Pogorui and Turbin (2002) and Pogorui (2003), the methodology of Markov evolutions is used for the study of a two-phase system with two reservoirs. This method does not need the assumption of the instantaneous addition of reserves in reservoirs. In Pogorui (2004a), system effectiveness is studied by using the phase merging scheme (Korolyuk and Swishchuk 1995b; Korolyuk and Korolyuk 1999; Korolyuk and Limnios 2005). So, in this section, the operation of the system is modeled by a random evolution in an alternating semi-Markov medium.

4.4.1. *Estimation of stationary efficiency of one-phase system with a reservoir*

Consider a system that consists of a supply line or flow A, a reservoir (accumulator, container or buffer) B and a customer C.

The supply line A is unreliable, with ON-OFF cycles: the time of non-failure operation (ON period) τ_1 of A, is a random variable with a general distribution function $G_1(t)$, and the time of failure operation (OFF period) τ_0 of A can be represented by a random variable with a general distribution function $G_0(t)$.

Let us denote by V the volume (or capacity) of the reservoir B. The system operates in the following manner:

1) If the supply line A is ON and the reservoir B is full, then the product goes from A to C at rate a_0.

2) If the supply line A is ON and the reservoir B is not full, then the product goes from A to C at rate a_0. In addition, the product goes from A to B at rate b.

3) If the supply line A is OFF and the reservoir B is not empty, then the product goes from B to C at rate a_1.

4) If the reservoir B is full, then the supply line does not provide any product to the reservoir.

5) If the supply line A is OFF and the reservoir B is empty, then the customer C does not receive any product.

Denote by V_T, the product total amount that is received by the customer during time $[0, T]$.

The limit $K = \lim_{T\to\infty} \frac{V_T}{T}$, if it exists, will be called the **stationary average productivity** of the system.

Our purpose is to calculate K as a function of the following system parameters: a_0, a_1, b, V, and distributions $G_0(t)$, $G_1(t)$.

Let us consider the following random process $\{n(t)\}$:

$$n(t) = \begin{cases} 0, & \text{if the supply line } A \text{ is OFF;} \\ 1, & \text{if the supply line } A \text{ is ON.} \end{cases}$$

The process $n(t)$ is semi-Markov, with the phase space $\mathrm{E} = \{0, 1\}$. A sojourn time of $n(t)$ in state 0 (respectively, 1) is τ_0 (respectively, τ_1).

Let $v(t)$ be the volume of the product in the reservoir B at time t. We introduce the following functions on $W = \{0, 1\} \times [0, V]$:

$$C(w) = \begin{cases} b, & w = (1, v), \quad v < V, \\ -a_1, & w = (0, v), \quad v > 0, \\ 0, & \text{other cases.} \end{cases}$$

$$f(w) = \begin{cases} a_0, & w = (1, v), \\ a_1, & w = (0, v), \quad v > 0, \\ 0, & w = (0, 0). \end{cases}$$

We can see that $v(t)$ satisfies the following evolution equation:

$$\frac{dv(t)}{dt} = C(n(t), v(t)), \qquad v(t) = v_0 \in [0, V].$$

Let us consider the following bivariate random process $\xi(t) = (n(t), v(t))$. Then we have

$$V_T = \int_0^T f(\xi(t))\, dt$$

and

$$\lim_{T\to\infty} \int_0^T f(\xi(t))\, dt = K$$

provided that K exists.

If the random process $\xi(t)$ has the stationary distribution $\rho(\cdot)$, then according to the ergodic theory, we have

$$\lim_{T\to\infty} \frac{1}{T} \int_0^T f(\xi(t))\, dt = \int_W f(w)\, \rho(dw).$$

Hence, to compute the stationary productivity K of the system, it suffices to determine the stationary distribution $\xi(t)$.

We assume that $G_i(t)$ is non-degenerate and that $g_i(t) = \frac{dG_i(t)}{dt}$ and $m_i = \int_0^\infty t g_i(t) dt$, $i = 0, 1$ exists.

Let us state as

$$\hat{g}_i(s) = \int_0^\infty e^{st} g_i(t) dt, \qquad r_i(t) = \frac{g_i(t)}{1 - G_i(t)}, \quad i = 0, 1.$$

Suppose the following conditions are fulfilled:

F_1. *If* $(m_0\, a_1 - m_1\, b) \neq 0$, *then there exists* $s_0 \neq 0$ *such that*

$$\hat{g}_0(a_1\, s_0)\, \hat{g}_1(-b\, s_0) = 1, \qquad [4.83]$$

F_2.

$$I_0 = \int_0^\infty exp\left\{a_1 s_0 t - \int_0^t r_0(l)\, dl\right\} dt < \infty,$$

$$I_1 = \int_0^\infty exp\left\{-b\, s_0 t - \int_0^t r_1(l)\, dl\right\} dt < \infty.$$

In the case when $(m_0\, a_1 - m_1\, b) = 0$ *instead of* F_2, *we will assume that the following condition is satisfied:*

$F_2{}'$. *There exists*

$$m_0^{(2)} = \int_0^\infty t^2 g_0(t)dt, \; m_1^{(2)} = \int_0^\infty t^2 g_1(t)dt.$$

LEMMA 4.3.– If $m_0\, a_1 \neq m_1\, b$ and $l_1 < l_2$, $p_1 < p_2$, $\sigma_1 > 0$, $\sigma_2 > 0$ exist, such that $g_0(t) \geq \sigma_1, t \in [l_1,\, l_2]$, $g_1(t) \geq \sigma_2$, $t \in [p_1,\, p_2]$ and $b\, p_1 < a_1 l_2$, $a_1\, l_1 < b\, p_2$, then $s_0 \neq 0$ which satisfies condition F_1 exists.

PROOF.– Let us define $f\,(s) = \hat{g}_0\,(a_1\, s)\, \hat{g}_0\,(-b\, s)$. We have $f'\,(0) = (m_0\, a_1 - m_1\, b) \neq 0$ and $f\,(0) = 1$. If $(m_0\, a_1 - m_1\, b) < 0$, then

$$f\,(s) \geq \int_{l_1}^{l_2} e^{a_1 s\, t} g_0(t)dt \int_{p_1}^{p_2} e^{-b\, s_0\, t} g_1(t)dt$$

$$\geq \frac{\sigma_1 \sigma_2}{a_1\, b} \left(e^{a_1 l_2 s} - e^{a_1 l_1 s}\right) \left(e^{-b\, p_1 s} - e^{-b\, p_2 s}\right) \to +\infty \;\; as\; s \to +\infty.$$

The case $f'\,(0) = (m_0\, a_1 - m_1\, b) > 0$ can be proved in a similar manner.

REMARK 4.2.– *We should note that condition F_2 is a consequence of condition F_1, provided that $\delta_1, \delta_2 > 0$ and $T < \infty$ exist such that $r_i(t) \geq \delta_i$ for $t > T$, $i = 1, 2$.*

THEOREM 4.5.– (Pogorui 2011b) If $m_0\, a_1 \neq m_1\, b$ and $s_0 \neq 0$ exists, such that conditions F_1, F_2, are fulfilled, then the stationary distribution of $\xi(t)$, with the absolutely continuous part exists:

$$\rho\,(0, v) = c\, s_0 I_0\, e^{s_0 v}, \quad v > 0;$$

$$\rho\,(1, v) = c\, \hat{g}_0\,(a_1 s_0)\, s_0 I_1\, e^{s_0 v}, \quad v < V.$$

and the singular part (or atoms) $\rho\,[0, 0]$, $\rho\,[1, V]$ at points $(0, 0)$, $(1, V)$, respectively:

$$\rho\,[0, 0] = c\,(I_0 - m_0)\,, \quad \rho\,[1, V] = c\, \hat{g}_0\,(a_1 s)\,(m_1 - I_1)\, e^{s_0 V},$$

where

$$c = \left[I_0 e^{s_0 V} - I_1 \hat{g}_0\,(a_1 s_0) - m_0 + m_1 \hat{g}_0\,(a_1 s_0)\, e^{s_0 V}\right]^{-1}.$$

If $m_0\, a_1 = m_1\, b$ and instead of F_2, condition $F_2{}'$ is fulfilled, then the stationary distribution of $\xi(t)$ with the absolutely continuous part exists:

$$\rho\,(i, v) = c\, m_i, \quad i = 0, 1$$

and the singular part (or atoms) at points $(0, 0)$, $(1, V)$:

$$\rho\,[0, 0] = c\, a_1 \frac{m_0^2}{2}, \quad \rho\,[1, V] = c\, b \frac{m_1^2}{2},$$

where $c^{-1} = V\,(m_0 + m_1) + a_1 \frac{m_0^2}{2} + b \frac{m_1^2}{2}$.

PROOF.– Let us consider the three-variate process $\xi(t) = (\tau(t),\ n(t), v(t))$ in the phase space $Z = [0,\infty) \times \{0,1\} \times [0,V]$, where

$$\tau(t) = t - \sup\{u < t : x(u) \neq x(t)\}.$$

It is well known that the process $\xi(t)$ is a Markov process and its infinitesimal operator is of the following form:

$$A\varphi(\tau,n,v) = \frac{\partial}{\partial\tau}\varphi(\tau,n,v) + r_x(\tau)\left[P\varphi(0,n,v) - \varphi(\tau,n,v)\right]$$

$$+C(n,v)\frac{\partial}{\partial v}\varphi(\tau,n,v),$$

where the function $\varphi(\tau,n,v)$ is continuously differentiable with respect to τ and v. In addition, $\varphi(\tau,n,v)$ satisfies the boundary conditions $\varphi_\tau{}'(\tau,1,0) = \varphi_\tau{}'(\tau,0,V) = 0$. Here, $(\tau,n,v) \in Z$, $P\varphi(0,1,v) = \varphi(0,0,v)$, $P\varphi(0,0,v) = \varphi(0,1,v)$.

In more details, the operator A has the form

$$A\varphi(\tau,1,v) = \frac{\partial}{\partial\tau}\varphi(\tau,1,v) + r_1(\tau)\left[\varphi(0,0,v) - \varphi(\tau,1,v)\right]$$

$$+b\frac{\partial}{\partial v}\varphi(\tau,1,v), \quad 0 \le v \le V,$$

$$A\varphi(\tau,0,v) = \frac{\partial}{\partial\tau}\varphi(\tau,0,v) + r_0(\tau)\left[\varphi(0,1,v) - \varphi(\tau,0,v)\right]$$

$$-a_1\frac{\partial}{\partial v}\varphi(\tau,0,v), \quad 0 \le v \le V,$$

$$A\varphi(\tau,0,0) = \frac{\partial}{\partial\tau}\varphi(\tau,0,0) + r_0(\tau)\left[\varphi(0,1,0) - \varphi(\tau,0,0)\right],$$

$$A\varphi(\tau,1,V) = \frac{\partial}{\partial\tau}\varphi(\tau,1,V) + r_1(\tau)\left[\varphi(0,0,V) - \varphi(\tau,1,V)\right],$$

$$\varphi_\tau{}'(\tau,1,0) = \varphi_\tau{}'(\tau,0,V) = 0.$$

If the stationary distribution $\rho(\cdot)$ of $\xi(t)$ exists, then for any function $\varphi(\cdot)$ from the domain of A, we have

$$\int_Z A\varphi(z)\,\rho(dz) = 0. \qquad [4.84]$$

The analysis of the properties of the process $\xi(t)$ leads to the conclusion that the stationary distribution $\rho(\cdot)$ has singularities at points $(\tau,0,0)$ and $(\tau,1,V)$, which are denoted by $\rho[\tau,0,0]$ and $\rho[\tau,1,V]$.

After the conversion of the operator A into the conjugate operator A^*, and using equation [4.84], we have

$$\frac{\partial}{\partial \tau}\rho(\tau,0,v)+r_0(\tau)\rho(\tau,0,v)-a_1\frac{\partial}{\partial v}\rho(\tau,0,v)=0, \quad [4.85]$$

$$\frac{\partial}{\partial \tau}\rho(\tau,1,v)+r_1(\tau)\rho(\tau,1,v)+b\frac{\partial}{\partial v}\rho(\tau,1,v)=0. \quad [4.86]$$

$$\int_0^\infty r_1(\tau)\rho(\tau,1,v)\,d\tau=\rho(0,0,v), \quad [4.87]$$

$$\int_0^\infty r_0(\tau)\rho(\tau,0,v)\,d\tau=\rho(0,1,v), \quad [4.88]$$

$$\rho(\infty,0,v)=\rho(\infty,1,v)=0,$$

$$\frac{\partial}{\partial \tau}\rho[\tau,0,0]+r_0(\tau)\rho[\tau,0,0]-a_1\rho(\tau,0,0+)=0, \quad [4.89]$$

$$\frac{\partial}{\partial \tau}\rho[\tau,1,V]+r_1(\tau)\rho[\tau,1,V]-b\rho(\tau,1,V-)=0, \quad [4.90]$$

$$\rho[\infty,1,V]=\rho[\infty,0,0]=\rho[0,1,V]=\rho[0,0,0]=0.$$

Now, by considering the boundary conditions, we obtain

$$\int_0^\infty r_0(\tau)\rho[\tau,0,0]\,d\tau=b\int_0^\infty \rho(\tau,1,0+)\,d\tau, \quad [4.91]$$

$$\int_0^\infty r_1(\tau)\rho[\tau,1,V]\,d\tau=a_1\int_0^\infty \rho(\tau,0,V-)\,d\tau. \quad [4.92]$$

Solving equations [4.85] and [4.86] gives for $f_i\in C^1$, $i=0,1$,

$$\rho(\tau,0,v)=f_0(v+a_1\tau)\,e^{-\int_0^\tau r_0(t)dt}, \quad [4.93]$$

$$\rho(\tau,1,v)=f_1(v-b\tau)\,e^{-\int_0^\tau r_1(t)dt}. \quad [4.94]$$

Substituting [4.93] and [4.94] into [4.87] and [4.88], we obtain

$$\int_0^\infty f_0(v+a_1t)\,e^{-\int_0^t r_0(s)ds}r_0(t)dt=f_1(v), \quad [4.95]$$

$$\int_0^\infty f_1(v-bt)\,e^{-\int_0^t r_1(s)ds}r_1(t)dt=f_0(v). \quad [4.96]$$

It follows from equations [4.95] and [4.96], for $i = 0, 1$, that

$$\int_0^\infty \int_0^\infty f_i\left(v - bt + a_1 l\right) e^{-\int_0^t r_1(s)ds} e^{-\int_0^l r_0(s)ds} r_1(t) r_0\left(l\right) dt dl = f_i\left(v\right). \qquad [4.97]$$

By combining equations [4.95]–[4.97], we will find the function $f_i\left(v\right)$ in the following form:

$$f_i\left(v\right) = c_i e^{sv}, \quad c_i > 0, \quad i = 0, 1. \qquad [4.98]$$

Next, substituting [4.93] into [4.92], and since $e^{-\int_0^t r_i(s)ds} = 1 - G_i(t)$, we have

$$\hat{g}_0\left(a_1 s\right) \hat{g}_1\left(-bs\right) = 1.$$

It follows from condition F_1 and equations [4.93]–[4.96] that

$$\rho\left(\tau, 0, v\right) = c_0 e^{s_0 v + a_1 s_0 \tau} e^{-\int_0^\tau r_0(t)dt}, \qquad [4.99]$$

$$\rho\left(\tau, 1, v\right) = c_0 \hat{g}_0\left(a_1 s\right) e^{s_0 v - b s_0 \tau} e^{-\int_0^\tau r_1(t)dt}. \qquad [4.100]$$

It is not hard to find that if $m_0\, a_1 \neq m_1\, b$, then only $s_0 \neq 0$ satisfies condition F_1. In addition, it also satisfies equations [4.91] and [4.92]. In particular cases, when $m_0\, a_1 = m_1\, b$, equations [4.91] and [4.92] have the solution $s_0 = 0$.

Substituting [4.99] and [4.100] into [4.89] and [4.90], respectively, we have

$$\rho\left[\tau, 0, 0\right] = c_0 a_1\, e^{-\int_0^\tau r_0(s)ds} \int_0^\tau e^{a_1 s_0 t} dt \qquad [4.101]$$

$$\rho\left[\tau, 1, V\right] = c_0 \hat{g}_0(a_1\, s_0) b\, e^{s_0 V}\, e^{-\int_0^\tau r_1(s)ds} \int_0^\tau e^{-b s_0 t} dt. \qquad [4.102]$$

The normalizing constant c_0 of equations [4.99]–[4.102] can be obtained from the condition

$$\sum_{i=0}^{1} \int_0^V \int_0^\infty \rho\left(\tau, i, v\right) d\tau dv + \int_0^\infty \rho\left[\tau, 0, 0\right] d\tau + \int_0^\infty \rho\left[\tau, 1, V\right] d\tau = 1.$$

To complete the proof, note that

$$\rho\left(i, v\right) = \int_0^\infty \rho\left(\tau, i, v\right) d\tau, \quad i = 0, 1,$$

$$\rho\left[0, 0\right] = \int_0^\infty \rho\left[\tau, 0, 0\right] d\tau, \qquad \rho\left[1, V\right] = \int_0^\infty \rho\left[\tau, 1, V\right] d\tau,$$

It follows from theorem 4.5 that if $m_0\, a_1 \neq m_1\, b$, then the stationary average productivity of the system is of the following form:

$$K = c\left[a_0 I_0\left(e^{s_0 V} - 1\right) + a_1 \hat{g}_0\left(a_1 s_0\right) I_1\left(e^{s_0 V} - 1\right)\right.$$
$$\left. + a_0 \hat{g}_0\left(a_1 s_0\right)\left(m_1 - I_1\right) e^{s_0 V}\right],$$

where

$$c = \left[I_0 e^{s_0 V} - I_1 \hat{g}_0\left(a_1 s_0\right) - m_0 + m_1 \hat{g}_0\left(a_1 s_0\right) e^{s_0 V}\right]^{-1}.$$

In the case when $m_0\, a_1 = m_1\, b$, which we call the balance case, we have

$$K = c\left[\left(m_0 a_1 + m_1 b\right) V + a_0 b \frac{m_1^{(2)}}{2}\right],$$

where $c^{-1} = V\left(m_0 + m_1\right) + a_1 \frac{m_0^{(2)}}{2} + b\frac{m_1^{(2)}}{2}$.

EXAMPLE 4.4.– Let $g_0(t) = q^2 e^{-qt}$, $q > 0, t \geq 0$, $g_1(t) = p^2 t e^{-pt}$, $p > 0$.

In this case, equation [4.83] has the form

$$\hat{g}_0\left(a_1 s_0\right) \hat{g}_1\left(-b s_0\right) = \left(\frac{q}{q - a_1 s_0}\right)^2 \left(\frac{p}{p + b s_0}\right)^2 = 1, \qquad [4.103]$$

with the conditions

$$\left(a_1\, s_0\right) < q, \quad b\, s_0 > -p. \qquad [4.104]$$

By solving equation [4.103], we obtain the following solutions that satisfy conditions [4.104], i.e. $s_0{}' = 0$ and $s_0{}'' = \frac{qb - a_1 p}{a_1 b}$.

Since $r_0(t) = \frac{q^2 t}{1+qt} \to q > 0$ as $t \to \infty$ and $r_1(t) = \frac{p^2 t}{1+pt} \to p > 0$ as $t \to \infty$, it follows from remark 4.2, that the integrals I_0, I_1 do exist.

Hence, we can apply theorem 4.5.

If we have the balance case, say $q\, b = a_1\, p$, then

$$K = \frac{2V\left(a_1 p^{-1} + b q^{-1}\right) + 3 a_0 b q^{-2}}{2V\left(p^{-1} + q^{-1}\right) + 3 a_1 p^{-2} + 3 b q^{-2}}.$$

In the unbalance case, say $q\, b \neq a_1\, p$ and $s_0{}'' = \frac{qb - a_1 p}{a_1 b}$, we have

$$K = c\Bigg[\left(a_0 I_0 + a_1 \hat{g}_0\left(\frac{qb - a_1 p}{b}\right) I_1\right)\left(e^{\frac{qb - a_1 p}{a_1 b} V} - 1\right)$$

$$+a_0 \hat{g}_0 \left(a_1 s_0''\right) \left(m_1 - I_1\right) e^{s_0'' V} \Bigg],$$

where

$$c = \left[I_0 e^{\frac{qb - a_1 p}{a_1 b} V} - I_1 \hat{g}_0 \left(\frac{qb - a_1 p}{b} \right) - \frac{1}{p} + \frac{1}{q} \hat{g}_0 \left(\frac{qb - a_1 p}{b} \right) e^{\frac{qb - a_1 p}{a_1 b} V} \right]^{-1},$$

and

$$\hat{g}_0 \left(\lambda\right) = \left(\frac{q}{q + \lambda} \right)^2,$$

$$I_0 = \int_0^\infty \left(1 + qt\right) e^{-\frac{a_1 p}{b} t} dt = \frac{qb^2 + a_1 pb}{a_1^2 p^2},$$

$$I_1 = \int_0^\infty \left(1 + pt\right) e^{-\frac{bp}{a_1} t} dt = \frac{pa_1^2 + a_1 qb}{b^2 q^2}.$$

4.4.2. *Estimation of stationary efficiency of a production system with two unreliable supply lines*

Now, we will deal with a system consisting of two unreliable supply lines A_1 and A_2, a customer C, and two reservoirs B_1 and B_2 located between A_1 and A_2, and also between A_2 and C, respectively. The ON time period of each supply line A_i is a random variable distributed exponentially with a parameter $\lambda_i > 0$ and the OFF periods are random variables distributed exponentially with parameter $\mu_i > 0$.

The system functions as follows:

1) If the supply lines A_1 and A_2 are ON and the reservoirs B_1 and B_2 are not full, then the flow (or product) arrives from A_1 to A_2 at rate a_0+b_2 and from A_2 to C at rate a_0. In addition, the reservoir B_1 is receiving a flow from A_1 at rate b_1, and the reservoir B_2 is receiving a flow from A_2 at rate b_2. If the reservoir B_1 is full and B_2 is not full, then the flow goes from A_1 to A_2 and from A_2 to C at rate a_0, and to the reservoir B_2 at rate b_2 (there is no flow to reservoir B_1). If B_2 is full and B_1 is not full, then the flow goes from A_1 to A_2 and from A_2 to C at rate a_0, and to the reservoir B_1 at rate b_1 (there is no flow going into reservoir B_2). If B_1 and B_2 are full, then the flow goes from A_1 to A_2, and from A_2 to C at rate a_0.

2) If A_1 is OFF, reservoir B_1 is not empty, and the reservoir B_2 is not full, then the flow goes from the non-empty B_1 to A_2 at rate $a_1 > b_2$, from A_2 to the reservoir B_2 at rate b_2, and to C at rate $a_1 - b_2$. If the reservoir B_2 is full, then the flow goes from A_2 to C at rate a_1, and there is no flow going into B_2.

3) If A_2 is OFF, then the flow goes from the non-empty reservoir B_2 to the customer C at rate a_2. If B_2 is empty, then the customer does not receive any flow or product. The supply line A_1 functions according to items 1 and 2, with the difference that the product does not arrive from A_1 to A_2.

As mentioned before, let V_T be the volume of the product received by the customer during an interval of time $[0, T]$. The limit $K = \lim_{T\to\infty} \frac{V_T}{T}$ (if any) is called the stationary average productivity of the system. Our purpose is to calculate K as a function of the following system parameters: a_1, a_2, b_1, $b_2, \mu_1, \mu_2, \lambda_1, \lambda_2$, V_1, V_2.

Let us introduce the following random process:

$$\chi(t) = \begin{cases} 11; & \text{supply lines } A_1 \text{ and } A_2 \text{ are ON;} \\ 10; & A_1 \text{ is ON and } A_2 \text{ is OFF;} \\ 01; & A_1 \text{ is OFF and } A_2 \text{ is ON;} \\ 00; & \text{supply lines } A_1 \text{ and } A_2 \text{ are OFF,} \end{cases}$$

at time t. The process $\{\chi(t)\}$ is a Markov process.

Denote by $v_i(t)$, the volume (or amount) of the product in B_i at time t, and by V_i, the volume or capacity of the reservoir B_i, $i = 1, 2$.

Now, consider the three-variate process $\zeta(t) = \{\chi(t), v_1(t), v_2(t)\}$. It is easily seen that $\zeta(t)$ is a Markov process with the following phase space:

$$X = \{(ij, v_1, v_2), \; i, j \in \{0, 1\}, \; 0 \le v_1 \le V_1, \; 0 \le v_2 \le V_2\}.$$

Let us define the following function on X:

$$f(x) = \begin{cases} 0, & \text{if } x = (i0, v_1, 0), \; i \in \{0,1\}, 0 \le v_1 \le V_1; \\ a_2, & \text{if } x = (i0, v_1, v_2), \; i \in \{0,1\}, 0 \le v_1 \le V_1, v_2 > 0; \\ a_0, & \text{if } x = (11, v_1, v_2); \\ a_1 - b_2, & \text{if } x = (01, v_1, v_2), v_1 > 0, v_2 < V_2; \\ a_1, & \text{if } x = (01, v_1, V_2), v_1 > 0. \end{cases}$$

Then, we have

$$V_T = \int_0^T f(\zeta(t))dt$$

and

$$\lim_{T\to\infty} \frac{V_T}{T} = \lim_{T\to\infty} \frac{1}{T} \int_0^T f(\zeta(t))dt = K.$$

The process $\zeta(t)$ is uniformly ergodic and we denote by $\rho(\cdot)$, its stationary distribution. According to the ergodic theory, we have

$$K = \lim_{T\to\infty} \frac{1}{T} \int_0^T f(\zeta(t))dt = \int_X f(x)\,\rho(dx).$$

Thus, to compute the stationary productivity K of the system it suffices to calculate $\rho(\cdot)$.

Let us introduce the following functions on the space X:

$$C_1(x, v_1, v_2) = \begin{cases} 0, & \text{if } x = (0,0) \text{ or } x = (1,1), v_1 = V_1; \\ b_1, & \text{if } x = (1,i), i \in \{0,1\}, v_1 < V_1; \\ -a_1, & \text{if } x = (0,1), v_1 > 0, \end{cases}$$

$$C_2(x, v_1, v_2) = \begin{cases} b_2, & \text{if } x = (1,1) \text{ or } x = (0,1), v_1 > 0, v_2 < V_2; \\ -a_2, & \text{if } x = (0,1), v_1 = 0, v_2 > 0 \text{ or } x = (0,0), v_2 > 0; \\ 0, & \text{other cases.} \end{cases}$$

It is easily verified that

$$\begin{cases} \dfrac{dv_1(t)}{dt} = C_1(\chi(t), v_1(t), v_2(t)); \\ \dfrac{dv_2(t)}{dt} = C_2(\chi(t), v_1(t), v_2(t)), \end{cases}$$

i.e. $\gamma(t) = (v_1(t), v_2(t))$ is a stochastic transport process (Korolyuk 1993; Korolyuk and Limnios 2005).

The infinitesimal operator of the process $\zeta(t) = (\chi(t), v_1(t), v_2(t))$ is as follows (Corlat *et al.* 1991; Korolyuk and Turbin 1982):

$$A\varphi(x, v_1, v_2) = C_1(x, v_1, v_2)\frac{\partial}{\partial v_1}\varphi(x, v_1, v_2)$$

$$+C_2(x, v_1, v_2)\frac{\partial}{\partial v_2}\varphi(x, v_1, v_2) + \frac{f_x}{1-F_x}(P-I)\varphi(x, v_1, v_2),$$

where F_x is the cdf and f_x is the pdf of a sojourn time of the process $\chi(t)$ at state $x \in \{11, 10, 01, 00\}$, P is the operator of the transition probabilities of $\chi(t)$ and I is a unit operator. By direct calculation, we obtain

$$\frac{f_{00}}{1-F_{00}} = \mu_1 + \mu_2, \qquad \frac{f_{01}}{1-F_{01}} = \mu_1 + \lambda_2,$$

$$\frac{f_{10}}{1-F_{10}} = \mu_2 + \lambda_1, \qquad \frac{f_{11}}{1-F_{11}} = \lambda_1 + \lambda_2. \qquad [4.105]$$

And the matrix P is as follows:

$$P = \begin{pmatrix} 0 & \frac{\mu_2}{\mu_2+\mu_1} & \frac{\mu_1}{\mu_2+\mu_1} & 0 \\ \frac{\lambda_2}{\lambda_2+\mu_1} & 0 & 0 & \frac{\mu_1}{\lambda_2+\mu_1} \\ \frac{\lambda_1}{\mu_2+\lambda_1} & 0 & 0 & \frac{\mu_2}{\mu_2+\lambda_1} \\ 0 & \frac{\lambda_1}{\lambda_1+\lambda_2} & \frac{\lambda_2}{\lambda_1+\lambda_2} & 0 \end{pmatrix} \qquad [4.106]$$

We use the following ordering in the binary system: $00 \Leftrightarrow 0$, $01 \Leftrightarrow 1$, $10 \Leftrightarrow 2$ and $11 \Leftrightarrow 3$.

The points of the phase space X, at which the functions $C_1(\cdot)$ and $C_2(\cdot)$ are simultaneously equal to zero, are atoms of the stationary distribution. We denote these atoms by $\rho[\cdot]$. The points at which the function $C_1(\cdot)$ or function $C_2(\cdot)$ (but not both) are equal to zero, are said to be semiatoms, and we denote them by $\rho[\cdot)$. At other points of the space X, the measure $\rho(\cdot)$ is absolutely continuous with respect to the Lebesgue measure on $[0, V_1] \times [0, V_2]$.

By obtaining the adjoint operator A^* corresponding to the operator A, and using the fact that $A^*\rho = 0$, we have

$$\begin{aligned} & a_1 \frac{\partial}{\partial v_1}\rho(01,v_1,v_2) - b_2\frac{\partial}{\partial v_2}\rho(01,v_1,v_2) + \mu_2\rho(00,v_1,v_2) + \lambda_1\rho(11,v_1,v_2) \\ & - (\lambda_2+\mu_1)\rho(01,v_1,v_2) = 0; \\ & a_2\frac{\partial}{\partial v_2}\rho(00,v_1,v_2) + \lambda_1\rho(10,v_1,v_2) + \lambda_2\rho(01,v_1,v_2) \\ & - (\mu_1+\mu_2)\rho(00,v_1,v_2) = 0; \\ & -b_1\frac{\partial}{\partial v_1}\rho(10,v_1,v_2) + a_2\frac{\partial}{\partial v_2}\rho(10,v_1,v_2) + \mu_1\rho(00,v_1,v_2) + \lambda_2\rho(11,v_1,v_2) \\ & - (\mu_2+\lambda_1)\rho(10,v_1,v_2) = 0; \\ & -b_1\frac{\partial}{\partial v_1}\rho(11,v_1,v_2) - b_2\frac{\partial}{\partial v_2}\rho(11,v_1,v_2) + \mu_1\rho(01,v_1,v_2) + \mu_2\rho(10,v_1,v_2) \\ & - (\lambda_1+\lambda_2)\rho(11,v_1,v_2) = 0. \end{aligned} \qquad [4.107]$$

We can also obtain equalities for atoms and semiatoms of the distribution $\rho(\cdot)$. However, it is difficult to solve such a set of equations. Therefore, we propose to consider a "corrected" model instead. Namely, it is assumed that if $v_1 > 0$, then $C_1(00, v_1, v_2) = -a_1$. That is, if both supply lines are OFF, then the product leaves B_1 at rate a_1, and neither the customer nor reservoir B_1 receive it. We denote the stationary productivity of such a system by K_0. It is obvious that $K_0 \leq K$.

COMMENT.– For a known value of K_0, the upper estimate can be easily obtained for K if we assume that, after leaving B_1 in a state $(00, v_1, v_2)$, $v_1>0$, the customer receives the product. We also note that the larger the value of $\max(\mu_1, \mu_2)$, the smaller the difference between K_0 and K.

The set of equations for the "corrected" model in comparison with equation [4.107], has one more item $(a_1 \frac{\partial}{\partial v_1} \rho(00, v_1, v_2))$ in the second equation, and it is given by

$$
\begin{aligned}
& a_1 \frac{\partial}{\partial v_1} \rho(01,v_1, v_2) - b_2 \frac{\partial}{\partial v_2} \rho(01,v_1, v_2) + \mu_2 \rho(00, v_1, v_2) \\
& +\lambda_1 \rho(11,v_1, v_2) - (\lambda_2 + \mu_1) \rho(01,v_1, v_2) = 0; \\
& a_1 \tfrac{\partial}{\partial v_1} \rho(00, v_1, v_2) + a_2 \tfrac{\partial}{\partial v_2} \rho(00, v_1, v_2) + \lambda_1 \rho(10,v_1, v_2) \\
& +\lambda_2 \rho(01,v_1, v_2) - (\mu_1 + \mu_2) \rho(00, v_1, v_2) = 0; \\
& -b_1 \frac{\partial}{\partial v_1} \rho(10,v_1, v_2) + a_2 \frac{\partial}{\partial v_2} \rho(10,v_1, v_2) + \mu_1 \rho(00, v_1, v_2) \\
& +\lambda_2 \rho(11,v_1, v_2) - (\mu_2 + \lambda_1) \rho(10,v_1, v_2) = 0; \\
& -b_1 \frac{\partial}{\partial v_1} \rho(11,v_1, v_2) - b_2 \frac{\partial}{\partial v_2} \rho(11,v_1, v_2) + \mu_1 \rho(01,v_1, v_2) \\
& +\mu_2 \rho(10,v_1, v_2) - (\lambda_1 + \lambda_2) \rho(11,v_1, v_2) = 0.
\end{aligned} \qquad [4.108]
$$

Hence, by solving [4.108], we have

$$
\begin{aligned}
\rho(00, v_1, v_2) &= c \exp\left(\left(\frac{\mu_1}{a_1} - \frac{\lambda_1}{b_1}\right) v_1 + \left(\frac{\mu_2}{a_2} - \frac{\lambda_2}{b_2}\right) v_2\right), \\
\rho(01,v_1, v_2) &= c \frac{a_2}{b_2} \exp\left(\left(\frac{\mu_1}{a_1} - \frac{\lambda_1}{b_1}\right) v_1 + \left(\frac{\mu_2}{a_2} - \frac{\lambda_2}{b_2}\right) v_2\right), \\
\rho(10,v_1, v_2) &= c \frac{a_1}{b_1} \exp\left(\left(\frac{\mu_1}{a_1} - \frac{\lambda_1}{b_1}\right) v_1 + \left(\frac{\mu_2}{a_2} - \frac{\lambda_2}{b_2}\right) v_2\right), \\
\rho(11,v_1, v_2) &= c \frac{a_1}{b_1} \frac{a_2}{b_2} \exp\left(\left(\frac{\mu_1}{a_1} - \frac{\lambda_1}{b_1}\right) v_1 + \left(\frac{\mu_2}{a_2} - \frac{\lambda_2}{b_2}\right) v_2\right),
\end{aligned} \qquad [4.109]
$$

where c is a normalization constant.

The set of equations for the singular part of the "corrected" system is as follows:

$$
\begin{aligned}
& -b_1 \frac{d}{dv_1} \rho[11,v_1, V_2) + \mu_1 \rho[01,v_1, V_2) - (\lambda_1 + \lambda_2) \rho[11,v_1, V_2) \\
& \qquad +b_2 \rho(11,v_1, V_2^-) = 0;
\end{aligned}
$$

$$
\begin{aligned}
&-b_2\frac{d}{dv_2}\rho\,[11,V_1,v_2)+\mu_2\rho\,[10,V_1,v_2)-(\lambda_1+\lambda_2)\,\rho\,[11,V_1,v_2)\\
&\qquad+b_1\rho\left(11,V_1^-,v_2\right)=0;\\
&a_1\frac{d}{dv_1}\rho\,[01,v_1,V_2)+\lambda_1\rho\,[11,v_1,V_2)-(\lambda_2+\mu_1)\,\rho\,[01,v_1,V_2)\\
&\qquad+b_2\rho\left(01,v_1,V_2^-\right)=0;\\
&a_2\frac{d}{dv_2}\rho\,[10,V_1,v_2)+\lambda_2\rho\,[11,V_1,v_2)-(\mu_2+\lambda_1)\,\rho\,[10,V_1,v_2)\\
&\qquad+b_1\rho\left(10,V_1^-,v_2\right)=0;\\
&\lambda_1\rho\,[10,V_1,v_2)-(\mu_1+\mu_2)\,\rho\,[00,V_1,v_2)=0;\\
&\lambda_2\rho\,[11,v_1,V_2)-a_2\rho\left(10,v_1,V_2^-\right)=0;\\
&\lambda_2\rho\,[01,v_1,V_2)+\mu_2\rho\,[00,V_1,v_2)-a_2\rho\left(00,v_1,V_2^-\right)=0;\\
&\lambda_1\rho\,[11,V_1,v_2)-a_1\rho\left(01,V_1^-,v_2\right)=0,
\end{aligned}
\qquad [4.110]
$$

where

$$
\rho\left(x,V_1^-,v_2\right)=\lim_{v_1\to V_1}\rho\left(x,v_1,v_2\right),\qquad \rho\left(x,v_1,V_2^-\right)=\lim_{v_2\to V_2}\rho\left(x,v_1,v_2\right).
$$

By considering [4.109], we have the following solution of [4.110]:

$$
\begin{aligned}
\rho\,[11,v_1,V_2)&=c\frac{a_1a_2}{\lambda_2b_1}\exp\left(\left(\frac{\mu_1}{a_1}-\frac{\lambda_1}{b_1}\right)v_1+\left(\frac{\mu_2}{a_2}-\frac{\lambda_2}{b_2}\right)V_2\right),\\
\rho\,[01,v_1,V_2)&=c\frac{a_2}{\lambda_2}\exp\left(\left(\frac{\mu_1}{a_1}-\frac{\lambda_1}{b_1}\right)v_1+\left(\frac{\mu_2}{a_2}-\frac{\lambda_2}{b_2}\right)V_2\right),\\
\rho\,[11,V_1,v_2)&=c\frac{a_1a_2}{\lambda_1b_2}\exp\left(\left(\frac{\mu_1}{a_1}-\frac{\lambda_1}{b_1}\right)V_1+\left(\frac{\mu_2}{a_2}-\frac{\lambda_2}{b_2}\right)v_2\right),\\
\rho\,[10,V_1,v_2)&=c\frac{a_1}{\lambda_1}\exp\left(\left(\frac{\mu_1}{a_1}-\frac{\lambda_1}{b_1}\right)V_1+\left(\frac{\mu_2}{a_2}-\frac{\lambda_2}{b_2}\right)v_2\right).
\end{aligned}
\qquad [4.111]
$$

We can write a system of equations for semiatoms, when the first reservoir is empty for the "corrected" system, as follows:

$$
\begin{aligned}
&\mu_1\rho\,[00,0,v_2)-b_1\rho\left(10,0^+,v_2\right)=0;\\
&\mu_1\rho\,[01,0,v_2)-b_1\rho\left(11,0^+,v_2\right)=0;\\
&a_2\frac{d}{dv_2}\rho\,[01,0,v_2)-(\mu_1+\lambda_2)\,\rho\,[01,0,v_2)+\mu_2\rho\,[00,0,v_2)\\
&\qquad+a_1\rho\,(01,0,v_2)=0;
\end{aligned}
\qquad [4.112]
$$

$$a_2 \frac{d}{dv_2} \rho [00, 0, v_2) - (\mu_1 + \mu_2) \rho [00, 0, v_2) + \lambda_2 \rho [01, 0, v_2)$$
$$+ a_1 \rho (00, 0, v_2) = 0$$

and when the second reservoir is empty:

$$\begin{aligned} &\mu_2 \rho [00, v_1, 0) - b_2 \rho \left(01, v_1, 0^+\right) = 0; \\ &-b_1 \frac{d}{dv_1} \rho [10, v_1, 0) - (\lambda_1 + \mu_2) \rho [10, v_1, 0) + \mu_1 \rho [00, v_1, 0) \\ &+ a_2 \rho \left(10, v_1, 0^+\right) = 0; \\ &\mu_2 \rho [10, v_1, 0) - \mu_2 \rho \left(11, v_1, 0^+\right) = 0; \\ &a_1 \frac{d}{dv_1} \rho [00, v_1, 0) - (\mu_1 + \mu_2) \rho [00, v_1, 0) + \lambda_1 \rho [10, v_1, 0) \\ &+ a_2 \rho \left(00, v_1, 0^+\right) = 0. \end{aligned} \qquad [4.113]$$

By solving [4.112] and [4.113], and considering [4.111], we have

$$\begin{aligned} \rho [00, 0, v_2) &= c \frac{a_1}{\mu_1} \exp \left(\left(\frac{\mu_2}{a_2} - \frac{\lambda_2}{b_2} \right) v_2 \right), \\ \rho [01, 0, v_2) &= c \frac{a_1 a_2}{\mu_1 b_2} \exp \left(\left(\frac{\mu_2}{a_2} - \frac{\lambda_2}{b_2} \right) v_2 \right), \\ \rho [00, v_1, 0) &= c \frac{a_2}{\mu_2} \exp \left(\left(\frac{\mu_1}{a_1} - \frac{\lambda_1}{b_1} \right) v_1 \right), \\ \rho [10, v_1, 0) &= c \frac{a_1 a_2}{\mu_2 b_1} \exp \left(\left(\frac{\mu_1}{a_1} - \frac{\lambda_1}{b_1} \right) v_1 \right). \end{aligned} \qquad [4.114]$$

Next, we will write the corresponding equations for atoms. If the reservoirs are full, then we have the system:

$$\begin{aligned} &-(\lambda_1 + \lambda_2) \rho [11, V_1, V_2] + b_2 \rho \left[11, V_1, V_2^-\right) + b_1 \rho \left[11, V_1^-, V_2\right) = 0; \\ &\lambda_1 \rho [11, V_1, V_2] - a_1 \rho \left[01, V_1^-, V_2\right) = 0; \\ &\lambda_2 \rho [11, V_1, V_2] - a_2 \rho \left[10, V_1, V_2^-\right) = 0. \end{aligned}$$

By applying the condition [4.111], we solve the system and it has the following solution:

$$\rho [11, V_1, V_2] = c \frac{a_1 a_2}{\lambda_1 \lambda_2} \exp \left(\left(\frac{\mu_1}{a_1} - \frac{\lambda_1}{b_1} \right) V_1 + \left(\frac{\mu_2}{a_2} - \frac{\lambda_2}{b_2} \right) V_2 \right). \qquad [4.115]$$

The system of equations for the atom $\rho[00,0,0]$ can be written as

$$-(\mu_1+\mu_2)\,\rho[00,0,0]+a_2\rho\left[00,0,0^+\right)+a_1\rho\left[00,0^+,0\right)=0;$$

$$\mu_1\rho[00,0,0]-b_1\rho\left[00,0^+,0\right)=0;$$

$$\mu_2\rho[00,0,0]-b_2\rho\left[00,0,0^+\right)=0.$$

By solving this system, after considering [4.114], the solution is given by

$$\rho[00,0,0]=\frac{a_1a_2}{\mu_1\mu_2}c. \qquad [4.116]$$

And, finally, we obtain the following system of equations for the atom $\rho[10,V_1,0]$:

$$-(\mu_1+\mu_2)\,\rho[10,V_1,0]+a_2\rho\left[10,V_1^-,0\right)+a_1\rho\left[10,V_1,0^+\right)=0;$$

$$\mu_1\rho[10,V_1,0]-b_1\rho\left[00,V_1^-,0\right)=0;$$

$$\mu_2\rho[10,V_1,0]-b_2\rho\left[11,V_1,0^+\right)=0.$$

By solving this system and considering [4.114], we have

$$\rho[10,V_1,0]=c\frac{a_1a_2}{\lambda_1\mu_2}\exp\left(\left(\frac{\mu_1}{a_1}-\frac{\lambda_1}{b_1}\right)V_1\right) \qquad [4.117]$$

The coefficient c is determined from the relation $\int_X \rho(dx)=1$. Namely,

$$\rho[00,0,0]+\rho[10,V_1,0]+\rho[11,V_1,V_2]+\int_0^{V_1}\rho[00,v_1,0)\,dv_1$$
$$+\int_0^{V_1}\rho[10,v_1,0)\,dv_1$$
$$+\int_0^{V_2}\rho[00,0,v_2)\,dv_2+\int_0^{V_2}\rho[01,0,v_2)\,dv_2+\int_0^{V_2}\rho[01,v_1,V_2)\,dv_1$$
$$+\int_0^{V_2}\rho[11,v_1,V_2)\,dv_1+\int_0^{V_2}\rho[10,V_1,v_2)\,dv_2+\int_0^{V_2}\rho[11,V_1,v_2)\,dv_2$$
$$+\int_0^{V_1}\int_0^{V_2}\rho(00,v_1,v_2)\,dv_2dv_1+\int_0^{V_1}\int_0^{V_2}\rho(00,v_1,v_2)\,dv_2dv_1$$
$$+\int_0^{V_1}\int_0^{V_2}\rho(01,v_1,v_2)\,dv_2dv_1+\int_0^{V_1}\int_0^{V_2}\rho(10,v_1,v_2)\,dv_2dv_1$$
$$+\int_0^{V_1}\int_0^{V_2}\rho(11,v_1,v_2)\,dv_2dv_1=1.$$

Substituting the values that have been found for ρ into this formula, we obtain

$$c^{-1} = \frac{b_1 b_2}{\mu_1 \mu_2} + \frac{b_1 b_2}{\mu_2} + \frac{a_1 a_2}{\mu_1 \mu_2} + \frac{\lambda_1 (b_2 + a_2)}{\mu_1 \mu_2} V_1 + \frac{b_2 + a_2}{\mu_2} V_1$$

$$+ \frac{\lambda_2 (b_1 + a_1)}{\mu_1 \mu_2} V_2 + \frac{b_1 + a_1}{\mu_1} V_2 + \left(\frac{\lambda_1 \lambda_2}{\mu_1 \mu_2} + \frac{\lambda_1}{\mu_1} + \frac{\lambda_2}{\mu_2} + 1 \right) V_1 V_2.$$

Substituting the value of c into [4.108], [4.110] and [4.113]–[4.116], we calculate

$$K_0 = \int_X f(x) \rho(dx) = a_0 \left(\rho[11, V_1, V_2] + \int_0^{V_2} \rho[11, v_1, V_2) \, dv_1 \right.$$

$$\left. + \int_0^{V_2} \rho[11, V_1, v_2) \, dv_2 + \int_0^{V_1} \int_0^{V_2} \rho(11, v_1, v_2) \, dv_2 dv_1 \right)$$

$$+ a_2 \left(\int_0^{V_2} \rho[00, 0, v_2) \, dv_2 + \int_0^{V_2} \rho[01, 0, v_2) \, dv_2 \right.$$

$$+ \int_0^{V_1} \int_0^{V_2} \rho(00, v_1, v_2) \, dv_2 dv_1$$

$$\left. + \int_0^{V_1} \int_0^{V_2} \rho(10, v_1, v_2) \, dv_2 dv_1 + \int_0^{V_2} \rho[10, V_1, v_2) \, dv_2 \right)$$

$$+ (a_1 - b_2) \int_0^{V_1} \int_0^{V_2} \rho(01, v_1, v_2) \, dv_2 dv_1 + a_1 \int_0^{V_2} \rho[01, v_1, V_2) \, dv_1.$$

Let us introduce the function:

$$f_1(x) = \begin{cases} f(x), & x \in X, \ x \neq (00, v_1, v_2), \\ a_1, & x = (00, v_1, v_2). \end{cases}$$

Then, it is easily seen that

$$\int_X f_1(x) \rho(dx) = K_0 + a_1 \frac{\lambda_1 \lambda_2}{\mu_1 \mu_2}.$$

Therefore, we have

$$K_0 \leq K \leq K_0 + a_1 \frac{\lambda_1 \lambda_2}{\mu_1 \mu_2}.$$

The approach that we use in this section can be applied to study the stationary efficiency of a production system with an arbitrary number of unreliable supply lines and reservoirs in between them.

5

One-dimensional Random Motions in Markov and Semi-Markov Media

In this chapter, several models of random evolutions which generalize the telegraph process of Goldstein–Kac are considered, and distributions of such evolutions are studied. In particular, section 5.1 is devoted to one-dimensional semi-Markov evolutions in a generalized Erlang environment. The selection of this distribution is due to the fact that, in this case, we are able to write a partial differential equation (PDE) of hyperbolic type for the density distribution of the particle position, which can be solved by a method described in section 5.1.2. In section 5.1.1, we obtain a hyperbolic differential equation for the pdf of a random motion with Erlang distributed velocity alternations.

In section 5.1.2, we describe the solution method for the PDE derived in section 5.1.1. The solution method generalizes the connection between holomorphic functions and harmonic functions through relationships between monogenic functions on a commutative algebra with associated PDEs

By using this method, in section 5.1.3 we have found the pdf of the random walk in fixed time described in section 5.1.1.

Section 5.2 is aimed at the study of fading evolutions, where, in contrast to the Goldstein–Kac model, a particle that is moving at variable speed slows down to zero as time goes to infinity, i.e. the particle slows down its chaotic movement up to the freezing point as $t \to \infty$. These processes simulate the motion of a particle under the influence of external forces, such as friction. The distribution of the limiting position of a fading evolution in the cases of Erlang-2 or uniform distributions for an interarrival time of the renewal process is obtained.

In section 5.3, we obtain a differential equation for the characteristic function of random jump motion on the line, where the direction alternations and random jumps occur according to the renewal epochs of the Erlang distribution.

Section 5.4 is devoted to the estimation of the number of level crossings by the telegraph process. By passing to the limit in Kac's condition, the estimation for a level crossing by the Wiener process is obtained.

5.1. One-dimensional semi-Markov evolutions with general Erlang sojourn times

In this section, we study a one-dimensional random motion with a general Erlang distribution for the sojourn times with velocities $v_1 > 0$ and $v_2 < 0$. This stochastic process generalizes the well-known Goldstein–Kac process because it considers the evolution in an Erlang (i.e. non-Markov) media. We should mention that many of the published papers on random motions make the typical assumption that motions are driven by a homogeneous Poisson process, so their processes are Markov (Di Crescenzo 2001; Orsingher and De Gregorio 2007; Pinsky 1991). Thus, their mathematical techniques cannot be extended to the non-Markov case in general. For instance, in the case of a non-Markov switching velocity process, it is not evident how to use the theory of infinitesimal operators and obtain Kolmogorov equations for the distribution of the random motion. However, in the specific case of the Erlang case, we can extend some of the same ideas by increasing the phase space, and as a consequence we can reduce the case of non-Markov switching to the well-known Markov case.

We also develop a novel method based on commutative algebras for solving this hyperbolic PDE and obtain the distribution of the particle position.

5.1.1. *Mathematical model*

We assume that the particle moves on the line $\mathbb{R}$ in the following manner: at each instant, it moves according to one of two velocities, namely $v_1 > 0$ or $v_2 < 0$. Starting at the position $x_0 \in \mathbb{R}$, the particle continues its motion with velocity $v_1 > 0$ during the random time $\tau_1 = \xi_{\lambda_1} + \ldots + \xi_{\lambda_n}$, $n \geq 1$, where $\{\xi_{\lambda_i}\}$ is a set of independent exponential random variables with parameters λ_i, respectively. Then the particle moves with velocity $v_2 < 0$ during the random time $\tau_2 = \xi_{\mu_1} + \ldots + \xi_{\mu_m}$, $m \geq 1$, where $\{\xi_{\mu_i}\}$ is a set of independent exponential random variables with parameters μ_i, respectively. Moreover, the particle moves with velocities $v_1 > 0$ and $v_2 < 0$, i.e. after an even renewal epoch, the particle has the velocity v_1, and after an odd renewal epoch it has the velocity v_2.

Thus, we study this one-dimensional random motion by assuming a general Erlang distribution for the sojourn times, meaning that the random variables τ_1 and τ_2 have general Erlang probability densities with Laplace transforms

$$\hat{f}_{\tau_1}(s) = \int_0^{\infty} e^{-st} f_{\tau_1}(t) dt = \prod_{i=1}^{n} \frac{\lambda_i}{s+\lambda_i};$$

$$\hat{f}_{\tau_2}(s) = \int_0^{\infty} e^{-st} f_{\tau_2}(t) dt = \prod_{i=1}^{m} \frac{\mu_i}{s+\mu_i}.$$

We will show that the movement of this particle can be determined in terms of a random evolution. Let us consider on the phase space $T = \{1, 2\}$ an alternating semi-Markov stochastic process $\eta(t), \ t \geq 0$ with the sojourn time τ_i, corresponding to the state $i \in T$, and transition probability matrix of its corresponding embedded Markov chain

$$P = \begin{pmatrix} 0 & 1 \\ 1 & 0 \end{pmatrix}.$$

Denote by $y(t), \ t \geq 0$ the particle position at time t. Now, consider the function C_1 on T as

$$C_1(y) = \begin{cases} v_1, & \text{if } y = 1, \\ v_2, & \text{if } y = 2. \end{cases} \qquad [5.1]$$

Then, the position of the particle at anytime t can be expressed as

$$y(t) = x_0 + \int_0^t C_1(\eta(s)) ds. \qquad [5.2]$$

Now, we show that the evolution $y(t)$ in semi-Markov media $\eta(t)$ can be reduced to a random evolution in a Markov media. Denote by $\{\xi(t), \ t \geq 0\}$ a Markov chain with the phase space $E = \{1, 2, \ldots, n, n+1, \ldots, n+m\}$ and the infinitesimal operator $Q = q[P - I]$, where

$$q = \text{diag}(\lambda_1, \lambda_2, \ldots, \lambda_n, \mu_1, \ldots, \mu_m)$$

is the intensities matrix and the transition probability matrix is given by

$$P = \begin{pmatrix} 0 & 1 & 0 & \ldots & 0 \\ 0 & 0 & 1 & \ldots & 0 \\ \vdots & \vdots & \vdots & \ddots & \vdots \\ 0 & 0 & 0 & \ldots & 1 \\ 1 & 0 & 0 & \ldots & 0 \end{pmatrix}.$$

We will assume that the initial distribution is $P\{\xi(0) = 1\} = 1$.

Now, let us introduce the following function C on E:

$$C(i) = \begin{cases} v_1, & \text{if } i = 1, \ldots, n, \\ v_2, & \text{if } i = n+1, \ldots, n+m. \end{cases}$$

Then the following equation

$$x(t) = x_0 + \int_0^t C(\xi(s))\, ds. \qquad [5.3]$$

determines the random evolution in the Markov medium $\{\xi(t),\ t \geq 0\}$.

The random processes $y(t)$ and $x(t)$ are stochastically equivalent and they model the particle motion evolution with velocities v_1 and v_2, and the random times between velocities switching have probability density function $f_{\tau_1}(t)$, if the velocity is v_1, and $f_{\tau_2}(t)$, when the velocity of the particle is v_2. Basically, we have reduced the semi-Markov case given by equation [5.2], to the Markov case given by equation [5.3]. So, we now need to study the stochastic process given by equation [5.3].

Let us consider the bivariate stochastic process $\zeta(t) = (x(t), \xi(t))$ with phase space $Z = \mathbb{R} \times E$, where $E = \{1, \ldots, n, n+1, \ldots, n+m\}$. It is well-known that the infinitesimal operator of $\zeta(t)$ is of the following form:

$$A\varphi(x,i) = C(x,i)\frac{\partial}{\partial x}\varphi(x,i) + q\,(P\varphi(x,i) - \varphi(x,i)),$$

where $\varphi \in D(A)$ = domain of operator A, $x \in R$, $i \in E$, and $C(x,i) = C(i)$ in this case. Let us write the operator A in more detail

$$A\varphi(x,1) = v_1\frac{\partial}{\partial x}\varphi(x,1) + \lambda_1(\varphi(x,2) - \varphi(x,1)),$$

$$A\varphi(x,2) = v_1\frac{\partial}{\partial x}\varphi(x,2) + \lambda_2(\varphi(x,3) - \varphi(x,2)),$$

$$\vdots$$

$$A\varphi(x,n) = v_1\frac{\partial}{\partial x}\varphi(x,n) + \lambda_n(\varphi(x,n+1) - \varphi(x,n)),$$

$$A\varphi(x,n+1) = v_2\frac{\partial}{\partial x}\varphi(x,n+1) + \mu_1(\varphi(x,n+2) - \varphi(x,n+1)),$$

$$\vdots$$

$$A\varphi(x,n+m) = v_2\frac{\partial}{\partial x}\varphi(x,n+m) + \mu_m(\varphi(x,1) - \varphi(x,n+m)).$$

Now, let us consider the density function

$$f(t, x, n)\, dx = P\{x \leq x(t) \leq x + dx, \zeta = n\}, \quad n \in E.$$

This function satisfies the first Kolmogorov equation, namely:

$$\frac{\partial f(t, x, n)}{\partial t} = Af(t, x, n). \qquad [5.4]$$

Equation [5.4] can be written in more detail as follows:

$$\begin{aligned}
\frac{\partial f(t, x, 1)}{\partial t} &= v_1 \frac{\partial}{\partial x} f(t, x, 1) + \lambda_1 (f(t, x, 2) - f(t, x, 1)), \\
\frac{\partial f(t, x, 2)}{\partial t} &= v_1 \frac{\partial}{\partial x} f(t, x, 2) + \lambda_2 (f(t, x, 3) - f(t, x, 2)), \\
&\dots \\
\frac{\partial f(t, x, n)}{\partial t} &= v_1 \frac{\partial}{\partial x} f(t, x, n) + \lambda_n (f(t, x, n+1) - f(t, x, n)), \\
\frac{\partial f(t, x, n+1)}{\partial t} &= v_2 \frac{\partial}{\partial x} f(t, x, n+1) \\
&\quad + \mu_1 (f(t, x, n+2) - f(t, x, n+1)), \\
&\vdots \\
\frac{\partial f(t, x, n+m)}{\partial t} &= v_2 \frac{\partial}{\partial x} f(t, x, n+m) \\
&\quad + \mu_m (f(t, x, 1) - f(t, x, n+m)).
\end{aligned} \qquad [5.5]$$

The set of equations [5.5] can be represented as

$$M\mathbf{f} = 0,$$

where $\mathbf{f} = (f(t, x, 1), f(t, x, 2), \dots, f(t, x, n+m))$, and M is equal to

$$\begin{pmatrix}
\frac{\partial}{\partial t} - v_1 \frac{\partial}{\partial x} + \lambda_1 & -\lambda_1 & 0 & \dots & 0 & 0 \\
0 & \frac{\partial}{\partial t} - v_1 \frac{\partial}{\partial x} + \lambda_2 & -\lambda_2 & \dots & 0 & 0 \\
\vdots & \vdots & \vdots & \ddots & \vdots & \vdots \\
0 & 0 & 0 & \dots & \frac{\partial}{\partial t} - v_2 \frac{\partial}{\partial x} + \mu_{m-1} & -\mu_{n+m-1} \\
-\mu_{n+m} & 0 & 0 & \dots & 0 & \frac{\partial}{\partial t} - v_2 \frac{\partial}{\partial x} + \mu_m
\end{pmatrix}$$

The probability density function of a particle position at time t is given by

$$f(t, x) = \sum_{i=1}^{n+m} f(t, x, i).$$

LEMMA 5.1.– The function $f(t,x)$ satisfies the following equation:

$$\left[\prod_{i=1}^{n}\left(\frac{\partial}{\partial t}-v_1\frac{\partial}{\partial x}+\lambda_i\right)\prod_{j=1}^{m}\left(\frac{\partial}{\partial t}-v_2\frac{\partial}{\partial x}+\mu_j\right)-\prod_{i=1}^{n}\lambda_i\prod_{j=1}^{m}\mu_j\right]$$

$$\times f(t,x)=0. \qquad [5.6]$$

PROOF.– It is well-known that $f(t,x)$ satisfies

$$(\det(M))\,f=0, \qquad [5.7]$$

It is easily seen that

$$\det(M)=\prod_{i=1}^{n}\left(\frac{\partial}{\partial t}-v_1\frac{\partial}{\partial x}+\lambda_i\right)\prod_{j=1}^{m}\left(\frac{\partial}{\partial t}-v_2\frac{\partial}{\partial x}+\mu_j\right)-\prod_{i=1}^{n}\lambda_i\prod_{j=1}^{m}\mu_j.$$

Since velocity v_i is finite for each time $t>0$, the process $\{x(t)\}$ is restricted to the interval $[v_2t,\ v_1t]$. The function $f(t,x)$ can be expressed as $f(t,x)=f_c(t,x)+f_s(t,x)$, where $f_c(t,x)$ is the absolute continuous part of the distribution of the process $x(t)$ with respect to the Lebesgue measure on the line, and $f_s(t,x)$ is the singular part of the distribution.

Now, let us consider $v_1=-v_2=v$, $n=m$, $\lambda_i=\mu_i=\lambda$.

LEMMA 5.2.– The absolute continuous part $f_c(t,x)$ is bounded for each $t>0$ and the singular part $f_s(t,x)$ is given by

$$f_s(t,x)=e^{-\lambda t}\left(\sum_{i=0}^{m-1}\frac{(\lambda t)^i}{i!}\right)\delta(tv-x). \qquad [5.8]$$

PROOF.– The singular part [5.8] states the fact that the particle starts at $x=0$ with velocity v, and before time t the velocity direction does not change.

Let us show that for $|x|<vt$ the distribution $f(t,x)$ has no singularity with respect to the Lebesgue measure on $\mathbb{R}$.

Denote by ν_k the random event "k changes of velocities occur".

For $\Delta x=[x,x+\Delta]$, where $\Delta>0$, consider the probability

$$P_{\overline{\nu}_0}\{x(t)\in\Delta x\}=\sum_{k\geq 1}P\{x(t)\in\Delta x,\ \nu_k\},$$

where $P_{\overline{\nu}_0}\{x(t) \in \Delta x\}$ is the probability that $x(t) \in \Delta x$ and at least one change of velocity occurs.

Now, we show that for all $t > 0$ there exists a constant $C_t < \infty$ such that

$$\sup_x \frac{P_{\overline{\nu}_0}\{x(t) \in \Delta x\}}{\Delta x} < C_t.$$

Denote by θ_k, $k \geq 1$ a time period between $(k-1)$th and kth changes of velocity direction. Recall that $\theta_k, \ k \geq 1$ are independent random variables with Erlang-m distribution.

It is easily seen that

$$\begin{aligned}
& P_{\overline{\nu}_0}\{x(t) \in \Delta x\} \\
& = \sum_{k \geq 1} P\left\{\sum_{i=1}^{k} (-1)^{i+1}\theta_i v + (-1)^k \left(t - \sum_{i=1}^{k}\theta_i\right) v \in \Delta x, \sum_{i=1}^{k}\theta_i < t\right\} \\
& = \sum_{k \geq 1} P\left\{\left(\sum_{i=1}^{k} (-1)^{i+1}\theta_i - (-1)^k \sum_{i=1}^{k}\theta_i\right) v \in \Delta x - (-1)^k vt, \sum_{i=1}^{k}\theta_i < t\right\} \\
& = \sum_{l \geq 0} P\left\{2v(\theta_1 + \theta_3 + \cdots + \theta_{2l+1}) \in \Delta x + vt, \sum_{i=1}^{2l+1}\theta_i < t\right\} \\
& \quad + \sum_{l \geq 0} P\left\{-2v(\theta_2 + \theta_4 + \cdots + \theta_{2l+2}) \in \Delta x - vt, \sum_{i=1}^{2l+2}\theta_i < t\right\} \\
& \leq \sup_{|x| \leq vt} \sum_{l \geq 0} P\{2v(\theta_1 + \theta_3 + \ldots + \theta_{2l+1}) \in \Delta x, \\
& \qquad 2v(\theta_2 + \theta_4 + \ldots + \theta_{2l}) < 2vt - x\} \\
& \quad + \sup_{|x| \leq vt} \sum_{l \geq 0} P\{-2v(\theta_2 + \theta_4 + \cdots + \theta_{2l+2}) \in \Delta x, \\
& \qquad 2v(\theta_1 + \theta_3 + \ldots + \theta_{2l+1}) < 2vt + x + \Delta\}.
\end{aligned}$$

Since $|x| \leq vt$, then for all $m \geq 1$ the density function $p_m(x, \lambda)$, corresponding to the Erlang-m distribution with parameter λ, satisfies $p_m(x, \lambda) \leq \lambda$, we have

$$\begin{aligned}
& \sum_{l \geq 1} P\{2v(\theta_1 + \theta_3 + \cdots + \theta_{2l+1}) \in \Delta x, 2v(\theta_2 + \theta_4 + \ldots + \theta_{2l}) < 2vt - x\} \\
& \leq \frac{\lambda\Delta}{2v} \sum_{l \geq 1} P\{\theta_2 + \theta_4 + \ldots + \theta_{2l} < 2t\}. \qquad [5.9]
\end{aligned}$$

Taking into account the distribution of θ_i, we have for $2lm + 1 > 2\lambda t$

$$P\{\theta_2 + \theta_4 + \ldots + \theta_{2l} < 2t\} \leq \left(e^{2\lambda t} - \sum_{i=0}^{2lm} \frac{(2\lambda t)^i}{i!} \right) e^{-2\lambda t}$$

$$\leq \frac{(2\lambda t)^{2lm+1} e^{-2\lambda t}}{2lm!(2lm + 1 - 2\lambda t)}.$$

Thus, considering [5.9], there exists a constant A_t such that

$$\sup_{|x| \leq vt} \sum_{l \geq 1} P\{-2v(\theta_1 + \theta_3 + \ldots + \theta_{2l-1}) \in \Delta x,$$

$$2v(\theta_2 + \theta_4 + \ldots + \theta_{2l}) < 2vt - x + \Delta\} \leq A_t \Delta.$$

Similarly, we can show that there exists a constant B_t such that

$$\sup_x \sum_{l \geq 1} P\{2v(\theta_2 + \theta_4 + \ldots + \theta_{2l}) \in \Delta x,$$

$$2v(\theta_1 + \theta_3 + \ldots + \theta_{2l-1}) < 2vt - x + \Delta\} \leq B_t \Delta.$$

The proof of lemma 5.2 concludes by defining $C_t = A_t + B_t$.

It is easily seen that if $v_1 = -v_2 = v$, $n = m$, $\lambda_i = \mu_i = \lambda$, then equation [5.6] is as follows:

$$\left(\frac{\partial}{\partial t} - v\frac{\partial}{\partial x} + \lambda \right)^m \left(\frac{\partial}{\partial t} + v\frac{\partial}{\partial x} + \lambda \right)^m f_c(t,x) - \lambda^{2m} f_c(t,x) = 0. \quad [5.10]$$

COROLLARY 5.1.– The absolute continuous part $f_c(t,x)$ satisfies equation [5.10] for $|x| < tv$.

Now, let us study the behavior of the continuous part $f_c(t,x)$ close to the lines $t = \pm\frac{x}{v}$. To avoid cumbersome expressions, we choose $v = 1$.

LEMMA 5.3.– For $m \geq 2$, we have

$$\lim_{\varepsilon \downarrow 0} \frac{P\{0 < t - x(t) < \varepsilon\}}{\varepsilon} = \frac{\lambda^m t^{m-1} e^{-\lambda t}}{2(m-1)!},$$

$$\lim_{\varepsilon \downarrow 0} \frac{P\{t + x(t) < \varepsilon\}}{\varepsilon} = 0.$$

PROOF.– It is easily verified that

$$P\{0 < t - x(t) \leq \varepsilon\} = P\left\{t - \frac{\varepsilon}{2} \leq \theta_1 < t\right\}$$

$$+ \int_0^t P\left\{\theta_3 \geq t - u, \theta_2 \leq \frac{\varepsilon}{2}, \theta_1 \in du\right\} + o(\varepsilon),$$

where θ_i, $i = 1, 2, 3$ are independent m-Erlang distributed random variables with parameter λ. Since $\int_0^t P\left\{\theta_3 \geq t - u, \theta_2 \leq \frac{\varepsilon}{2}, \theta_1 \in du\right\} = o(\varepsilon)$, then after applying the limit we obtain

$$\lim_{\varepsilon \downarrow 0} \frac{P\{0 < t - x(t) < \varepsilon\}}{\varepsilon} \qquad [5.11]$$

$$= \lim_{\varepsilon \downarrow 0} \frac{e^{-\lambda t}}{\varepsilon} \left(\left(\sum_{i=0}^{m-1} \frac{(\lambda t)^i}{i!} \right) - e^{\lambda \frac{\varepsilon}{2}} \left(\sum_{i=0}^{m-1} \frac{\left(\lambda \left(t - \frac{\varepsilon}{2}\right)\right)^i}{i!} \right) \right) = \frac{\lambda^m t^{m-1} e^{-\lambda t}}{2(m-1)!}.$$

Similarly, $P\{t + x(t) \leq \varepsilon\} = P\left\{\theta_2 \geq t - \frac{\varepsilon}{2}, \theta_1 \leq \frac{\varepsilon}{2}\right\} + o(\varepsilon)$, and, as is easily seen, $\lim_{\varepsilon \downarrow 0} \frac{P\{t + x(t) < \varepsilon\}}{\varepsilon} = 0$.

REMARK 5.1.– *The case where $m = 1$ will be considered below as an example.*

REMARK 5.2.– *We will seek for solutions of* [5.10] *among functions with the continuous part $f_c(t, x)$, and satisfying the limiting conditions*

$$\lim_{x \uparrow t} f_c(t, x) = \lim_{\varepsilon \downarrow 0} \frac{P\{0 < t - x(t) < \varepsilon\}}{\varepsilon},$$

$$\lim_{x \downarrow -t} f_c(t, x) = \lim_{\varepsilon \downarrow 0} \frac{P\{t + x(t) < \varepsilon\}}{\varepsilon}. \qquad [5.12]$$

By applying the transformation $f(t, x) = e^{\lambda t} g(t, x)$ and changing the variable $y = \frac{x}{v}$, we reduce [5.10] to

$$\left(\frac{\partial^2}{\partial t^2} - \frac{\partial^2}{\partial y^2} \right)^m g_c(t, y) - \lambda^{2m} g_c(t, y) = 0, \qquad [5.13]$$

with the singular part $g_s(t, y) = \left(\sum_{i=0}^{m-1} \frac{(\lambda t)^i}{i!} \right) \delta(t - y)$.

We can obtain solutions of equation [5.13] by investigating monogenic functions ranged in associated commutative algebras. We will develop this approach in the following section.

5.1.2. *Solution of PDEs with constant coefficients and derivability of functions ranged in commutative algebras*

In this section, a solution method for PDEs with constant coefficients, developed in (Pogorui 2007; Pogorui *et al.* 2014), is described.

It is well-known that the real and imaginary parts of any holomorphic function are harmonic functions of two variables. We extend this idea to finite-dimensional commutative algebras; that is, we prove that if some basis of a subspace of a commutative algebra satisfies a polynomial equation, then the components of a monogenic function on the subspace are solutions of the respective PDE.

Let A be a finite-dimensional commutative unitary algebra over $\mathbb{K} = \mathbb{R}$ (or $\mathbb{C}$). Assume that the set of vectors $\mathbf{e}_0, \mathbf{e}_1, \ldots, \mathbf{e}_n$ is a basis of A and that $\mathbf{e}_0$ is the unit element of this algebra. Suppose that B is an $(m+1)$-dimensional subspace of A with the basis $\mathbf{e}_0, \mathbf{e}_1, \ldots, \mathbf{e}_m$, $m \leq n$. Any function $\mathbf{f} : B \to A$ is of the form

$$\mathbf{f}(\mathbf{x}) = \sum_{k=0}^{n} \mathbf{e}_k u_k(\mathbf{x}),$$

where $u_k(\mathbf{x}) = u_k(x_0, x_1, \ldots, x_m)$ are $\mathbb{K}$-valued functions of $m+1$ variables $x_i \in \mathbb{K}$.

DEFINITION 5.1.– $\mathbf{f}(\mathbf{x})$ *is called differentiable at a point* $\mathbf{x}_0 \in B$ *if there exists a unique element* $\mathbf{f}'(\mathbf{x}_0) \in A$ *such that for any* $\mathbf{h} \in B$

$$\mathbf{f}'(\mathbf{x}_0)\mathbf{h} = \lim_{\varepsilon \to 0} \frac{\mathbf{f}(\mathbf{x}_0 + \varepsilon\mathbf{h}) - \mathbf{f}(\mathbf{x}_0)}{\varepsilon},$$

where $\varepsilon \in \mathbb{K}$, *and* $\mathbf{f}'(\mathbf{x}_0)\mathbf{h}$ *is the product of two elements* $\mathbf{f}'(\mathbf{x}_0)$ *and* $\mathbf{h}$ *of algebra* A.

REMARK 5.3.– *The existence of the limit at the right-hand side is equivalent to the condition: the Gateaux derivative along* $\mathbf{h}$ *is proportional to* $\mathbf{h}$. *Furthermore, the proportionality coefficient is the same for any direction* $\mathbf{h}$. *It is denoted by* $\mathbf{f}'(\mathbf{x}_0)$ *and called the derivative of* $\mathbf{f}$ *at the point* $\mathbf{x}_0$. *In the case where the algebra* A *is over* $\mathbb{C}$, *definition 5.1 implies Gateaux differentiability along complex lines. For* $A = B = \mathbb{C}$, *this definition is also equivalent to the (complex) differentiability of the complex function* $\mathbf{f}$, *and* $\mathbf{f}'(\mathbf{x}_0)$ *becomes the usual complex derivative.*

Thus, one can expect that the introduced functions inherit some important properties of holomorphic functions.

We say that $\mathbf{f} : B > A$ is differentiable (in B) or monogenic if it is differentiable at any point of B (Mel'nichenko and Plaksa 2008).

Definition 5.1 includes the existence of all $(m+1)$ partial derivatives. Indeed, it is enough to take $\mathbf{h} = \mathbf{e}_k$, $k \in \{0, 1, 2, \ldots, m\}$, and it implies

$$\frac{\partial}{\partial x_k}\mathbf{f}(\mathbf{x}_0) = \mathbf{e}_k \mathbf{f}'(\mathbf{x}_0)$$

for all k with the same coefficient $\mathbf{f}'(\mathbf{x}_0)$. Since $\mathbf{e}_0$ is the unit element, then $\frac{\partial}{\partial x_0}\mathbf{f}(\mathbf{x}_0) = \mathbf{f}'(\mathbf{x}_0)$.

THEOREM 5.1.– A function $\mathbf{f}(\mathbf{x}) = \sum_{k=0}^{n} \mathbf{e}_k u_k(\mathbf{x})$ is differentiable at $\mathbf{x}_0$ if and only if it has all partial derivatives proportional, with the same proportionality coefficient, to the respective basis vectors $\mathbf{e}_k$:

$$\frac{\partial \mathbf{f}(\mathbf{x}_0)}{\partial x_k} = \mathbf{e}_k \alpha(\mathbf{x}_0);$$

whenever it holds

$$\alpha(\mathbf{x}_0) = \mathbf{f}'(\mathbf{x}_0).$$

PROOF.– The necessity part is given already. Now we prove sufficiency.

Define $\mathbf{h} = \sum_{k=0}^{n} h_k \mathbf{e}_k$, then

$$\begin{aligned}
\mathbf{h}\mathbf{f}'(\mathbf{x}_0) &= \sum_{k=0}^{n} h_k \mathbf{e}_k \mathbf{f}'(\mathbf{x}_0) \\
&= h_0 \sum_{k=0}^{n} \mathbf{e}_k \frac{\partial u_k}{\partial x_0} + h_1 \mathbf{e}_1 \sum_{k=0}^{n} \mathbf{e}_k \frac{\partial u_k}{\partial x_0} + \ldots + h_n \mathbf{e}_n \sum_{k=0}^{n} \mathbf{e}_k \frac{\partial u_k}{\partial x_0} \\
&= h_0 \sum_{k=0}^{n} h_k \mathbf{e}_k \frac{\partial u_k}{\partial x_0} + h_1 \mathbf{e}_1 \sum_{k=0}^{n} h_k \mathbf{e}_k \frac{\partial u_k}{\partial x_1} + \ldots + h_n \mathbf{e}_n \sum_{k=0}^{n} \mathbf{e}_k \frac{\partial u_k}{\partial x_n} \\
&= \lim_{\varepsilon \to 0} \frac{\mathbf{f}(\mathbf{x}_0 + \varepsilon \mathbf{h}) - \mathbf{f}(\mathbf{x}_0)}{\varepsilon}.
\end{aligned}$$

Hence, $\mathbf{f}$ is differentiable at $\mathbf{x}_0$.

REMARK 5.4.– *The conditions* $\frac{\partial \mathbf{f}(\mathbf{x}_0)}{\partial x_k} = \mathbf{e}_k \frac{\partial \mathbf{f}(\mathbf{x}_0)}{\partial x_0}$, $k = 1, \ldots, m$ *can be written as*

$$\sum_{k=0}^{n} \mathbf{e}_k \frac{\partial u_k}{\partial x_j} = \mathbf{e}_j \sum_{k=0}^{n} \mathbf{e}_k \frac{\partial u_k}{\partial x_0}, \qquad [5.14]$$

$j = 1, \ldots, m$ *and* [5.14] *is said to be the conditions of the* ***Cauchy–Riemann type****. In the case where* $A = B = \mathbb{C}$, equations [5.14] *are actually the Cauchy–Riemann conditions.*

REMARK 5.5.– *It is easily verified that theorem 5.1 remains true even for B with basis $\mathbf{e}_1^{'}, \mathbf{e}_2^{'}, \ldots, \mathbf{e}_{m+1}^{'}$ that does not contain the unit element $\mathbf{e}_0$. In this case, the Cauchy–Riemann type of conditions* [5.14] *can be stated in the symmetric form as follows*:

$$\mathbf{e}_i \sum_{k=0}^{n} \mathbf{e}_k \frac{\partial u_k}{\partial x_j} = \mathbf{e}_j \sum_{k=0}^{n} \mathbf{e}_k \frac{\partial u_k}{\partial x_i}, \quad i, j = 1, \ldots, m.$$

For given integers $r, m \geq 1$, let

$$P(\xi_0, \xi_1, \ldots, \xi_m) = \sum_{i_0+i_1+\ldots+i_m=r} C_{i_0,i_1,\ldots,i_m} \xi_0^{i_0} \xi_1^{i_1} \ldots \xi_m^{i_m}, \qquad [5.15]$$

where $C_{i_0,i_1,\ldots,i_m} \in \mathbb{K}$ are constant coefficients.

Consider the following PDE

$$P(\partial_0, \partial_1, \ldots, \partial_m)\left[u(x_0, x_1, \ldots, x_m)\right] = 0, \qquad [5.16]$$

where $\partial_k = \frac{\partial}{\partial x_k}$.

THEOREM 5.2.– Let P be a polynomial as in equation [5.15]. Let a function $\mathbf{f} : B \to A$ and its derivatives $\mathbf{f}^{'}, \mathbf{f}^{''}, \ldots, \mathbf{f}^{r}$ be differentiable, $\mathbf{f}(\mathbf{x}) = \sum_{k=0}^{n} \mathbf{e}_k u_k(\mathbf{x})$. Assume that the basis $\mathbf{e}_0, \mathbf{e}_1, \ldots, \mathbf{e}_m$ of the subspace B of the algebra A is such that

$$P(\mathbf{e}_0, \mathbf{e}_1, \ldots, \mathbf{e}_m) = 0.$$

Then the functions $u_k(\mathbf{x}), \ k = 1, \ldots, n$ are solutions of equation [5.16].

PROOF.– It follows from the Cauchy–Riemann type condition [5.14] that

$$\frac{\partial^{i_n} \mathbf{f}(\mathbf{x})}{\partial x_k^{i_n}} = (\mathbf{e}_k)^{i_n} \frac{\partial \mathbf{f}(\mathbf{x})}{\partial x_0}.$$

This implies for $(i_0 + i_1 + \ldots + i_m) \leq r$ that

$$\frac{\partial^{i_0+i_1+\ldots+i_m}}{\partial x_0^{i_0} \partial x_1^{i_1} \ldots \partial x_m^{i_m}} \mathbf{f}(\mathbf{x}) = (\mathbf{e}_0)^{i_0} (\mathbf{e}_1)^{i_1} \ldots (\mathbf{e}_k)^{i_m} \frac{\partial^{i_0+i_1+\ldots+i_m}}{\partial x_0^{i_0+i_1+\ldots+i_m}} \mathbf{f}(\mathbf{x}).$$

Therefore, we obtain

$$\sum_{i_0+i_1+\ldots+i_m=r} C_{i_0,i_1,\ldots,i_m} \partial_0^{i_0} \partial_1^{i_1} \ldots \partial_m^{i_m} \mathbf{f}(\mathbf{x})$$

$$= \frac{\partial^r}{\partial x_0^r} \mathbf{f}(\mathbf{x}) \sum_{i_0+i_1+\ldots+i_m=r} C_{i_0,i_1,\ldots,i_m} (\mathbf{e}_0)^{i_0} (\mathbf{e}_1)^{i_1} \ldots (\mathbf{e}_k)^{i_m} = 0.$$

Hence, it follows that every component $u_k(x), \ k = 1, \ldots, n$, of the function $\mathbf{f}$ is a solution of equation [5.16].

REMARK 5.6.– *Similarly to theorem 5.1, theorem 5.2 remains true even if B does not contain the unit element $\mathbf{e}_0$.*

5.1.3. *Infinite-dimensional case*

Let us consider the case where A is an infinite-dimensional commutative unitary algebra over $\mathbb{K}$, with the basis $\mathbf{e}_m$, $m = 0, 1, 2, \ldots$, and B is an $(m+1)$-dimensional subspace of A (similarly to the finite-dimensional case).

Now, consider a function $\mathbf{f} : B \to A$ of the following form:

$$\mathbf{f}(\mathbf{x}) = \sum_{k=0}^{\infty} u_k(\mathbf{x})\,\mathbf{e}_k,$$

where $u_k(\mathbf{x}) = u_k(x_0, x_1, \ldots, x_m)$ are real or complex functions.

Similarly to theorem 5.1, it can be proved that the function $\mathbf{f}$ is monogenic if and only if

$$\frac{\partial \mathbf{f}(\mathbf{x}_0)}{\partial x_k} = \mathbf{e}_k \frac{\partial \mathbf{f}(\mathbf{x}_0)}{\partial x_0}, \quad k = 0, 1, 2, \ldots. \qquad [5.17]$$

Suppose that A is the $\mathbb{R}$ algebra of the Fourier series with the basis $\mathbf{e}_k = \cos(kt)$, $k = 0, 1, 2, \ldots$, and B is the following subspace $B = \{\sum_{k=0}^{m} x_k \mathbf{e}_k\}$, $x_k \in \mathbb{R}$. It is easily verified that

$$\mathbf{e}_k \mathbf{e}_m = \frac{1}{2}(\mathbf{e}_{k-m} + \mathbf{e}_{k+m}), \; k \geq m.$$

The Cauchy–Riemann type conditions for this case follow from [5.17]

$$\frac{\partial u_0}{\partial x_1} = \frac{1}{2}\frac{\partial u_1}{\partial x_0}, \quad \frac{\partial u_1}{\partial x_1} = \frac{\partial u_0}{\partial x_0} + \frac{1}{2}\frac{\partial u_2}{\partial x_0},$$

and

$$\frac{\partial u_k}{\partial x_1} = \frac{1}{2}\left(\frac{\partial u_{k-1}}{\partial x_0} + \frac{\partial u_{k+1}}{\partial x_0}\right), \quad k = 2, 3, \ldots.$$

Similarly,

$$\frac{\partial u_0}{\partial x_2} = \frac{1}{2}\frac{\partial u_2}{\partial x_0}, \quad \frac{\partial u_1}{\partial x_2} = \frac{1}{2}\left(\frac{\partial u_1}{\partial x_0} + \frac{\partial u_3}{\partial x_0}\right),$$

where $\frac{\partial u_2}{\partial x_2} = \frac{1}{2}\left(\frac{\partial u_0}{\partial x_0} + \frac{\partial u_4}{\partial x_0}\right)$, and $\frac{\partial u_k}{\partial x_2} = \frac{\partial u_{k-2}}{\partial x_0} + \frac{1}{2}\frac{\partial u_{k+2}}{\partial x_0}$, $k = 2, 3, \ldots$, and so on.

5.1.4. *The distribution of one-dimensional random evolutions in Erlang media*

Now, we will apply the previous methodology to obtain the pdf $f(t,x)$ of a particle position, whose random motion was described in section 5.1.1. To find the function $f(t,x)$, we solve equation [5.13] by using the algebraic approach presented in the previous section. We will assume, without loss of generality, that $\lambda = 1$.

Let A_0 be a $2m$-dimensional commutative algebra over $\mathbb{R}$, and assume that a set $e_0, e_1, \ldots, e_{2m-1}$ is a basis of A_0 with the following Cayley table:

$$e_i e_j = e_{i\oplus j},$$

where $i \oplus j = i + j \ (mod\ 2m)$.

The algebra A_0 has the matrix representation as follows:

$$e_k \to \mathbf{e}_k = P_1^k,$$

where $P_1 = [p_{ij}]_{2m\times 2m}$, $p_{i,i+1} = 1$ for $0 \le i \le 2m-1$, $p_{2m,0} = 1$ and $p_{ij} = 0$ for the rest of $i,\ j$.

Let us define for $l = 0, 1, 2, \ldots, 2m-1,\ k = 0, 1, 2, \ldots$

$$\boldsymbol{\tau}_0^l = \mathbf{e}_l,\ l = 0, 1, \ldots, 2m-1,$$

$$\boldsymbol{\tau}_1^l = \mathbf{e}_l i \sin s,\ l = 0, 1, \ldots, 2m-1,$$

$$\boldsymbol{\tau}_2^l = \mathbf{e}_l \cos s,\ l = 0, 1, 2, \ldots, 2m-1,$$

$$\boldsymbol{\tau}_{2k}^l = \mathbf{e}_l \cos ks,\ \boldsymbol{\tau}_{2k+1}^l = \mathbf{e}_l i \sin(k+1)s.$$

It is easily verified that $\boldsymbol{\tau}_{2n}^0 \boldsymbol{\tau}_{2k}^0 = \frac{1}{2}\left(\boldsymbol{\tau}_{2(n-k)}^0 + \boldsymbol{\tau}_{2(n+k)}^0\right),\ n \ge k,$

$$\boldsymbol{\tau}_{2n+1}^{l_1} \boldsymbol{\tau}_{2k+1}^{l_2} = \frac{1}{2}\left(\boldsymbol{\tau}_{2(n-k)}^{l_1\oplus l_2} - \boldsymbol{\tau}_{2(n+k)}^{l_1\oplus l_2}\right), n \ge k,$$

$$\boldsymbol{\tau}_{2n+1}^{l_1} \boldsymbol{\tau}_{2k}^{l_2} = \frac{1}{2}\left(\boldsymbol{\tau}_{2(n-k)+1}^{l_1\oplus l_2} + \boldsymbol{\tau}_{2(n+k)+1}^{l_1\oplus l_2}\right), n \ge k.$$

Consider the following algebra

$$A = \left\{ \sum_{k=0}^{+\infty} \sum_{l=0}^{2m-1} \left(a_{2k}^l \boldsymbol{\tau}_{2k}^l + a_{2k+1}^l \boldsymbol{\tau}_{2k+1}^l\right) \,\middle|\, a_j^l \in \mathbb{R} \right\},$$

where $\sum_{k=0}^{+\infty} \sum_{l=0}^{2m-1} \left(\left|a_{2k}^l\right| + \left|a_{2k+1}^l\right|\right) < +\infty$.

We should note that such Banach algebra was considered in Mel'nichenko and Plaksa (2008).

Let $B = \{a_0\boldsymbol{\tau}_1^1 + a_1\boldsymbol{\tau}_2^1 + a_2\boldsymbol{\tau}_0^0 \mid a_i \in \mathbb{R}\}$ be a subspace of algebra A.

Let us introduce the function $\mathbf{f}(t,y,z) = f(e_1(t\cos s + yi\sin s) + z)$ such that $\mathbf{f} : B \to A$, i.e.

$$\mathbf{f}(t,y,z) = \sum_{k=0}^{+\infty}\sum_{l=0}^{2m-1}\left(v_{2k}^l(t,y,z)\,\boldsymbol{\tau}_{2k}^l + v_{2k+1}^l(t,y,z)\,\boldsymbol{\tau}_{2k+1}^l\right).$$

Suppose that the function $\mathbf{f}$ is monogenic, i.e.

$$\frac{\partial}{\partial t}\mathbf{f} = e_1\cos s\,\frac{\partial}{\partial z}\mathbf{f} \qquad [5.18]$$

and

$$\frac{\partial}{\partial y}\mathbf{f} = e_1 i\sin s\,\frac{\partial}{\partial z}\mathbf{f}. \qquad [5.19]$$

Now, by introducing the function $\mathbf{f}(t,y,z) = e^z\mathbf{g}_c(t,y)$, we reduce equation [5.13] to the following equation:

$$\left(\frac{\partial^2}{\partial t^2} - \frac{\partial^2}{\partial y^2}\right)^m \mathbf{f}(t,y,z) - \frac{\partial^{2m}}{\partial z^{2m}}\mathbf{f}(t,y,z) = 0. \qquad [5.20]$$

In this case, all $v_{2k}^l(t,y,z)$ are solutions of equation [5.20]. Indeed,

$$\left(\frac{\partial^2}{\partial t^2} - \frac{\partial^2}{\partial y^2}\right)^m \mathbf{f} - \frac{\partial^{2m}}{\partial z^{2m}}\mathbf{f} = e_1^{2m}\left(\cos^2 s - (i\sin s)^2\right)^m - 1 = 0.$$

We will denote the element e_1 as $\mathbf{e}$, and we will seek a solution for equation [5.20] in the following form:

$$\mathbf{g}_c(\mathbf{e}(t\cos s + yi\sin s)) = e^{\mathbf{e}(t\cos s + yi\sin s)}$$

Since $\mathbf{f}(\mathbf{e}(t\cos s + yi\sin s) + z) = \mathbf{g}_c(\mathbf{e}(t\cos s + yi\sin s))e^z$, we have

$$v_k^l(t,y,z) = u_k^l(t,y)\,e^z,\ l = 0,1,\ldots,2m-1,\ k = 0,1,2,\ldots,$$

where $\mathbf{g}_c(t,y) = \sum_{k=0}^{+\infty}\sum_{l=0}^{2m-1}\left(u_{2k}^l(t,y)\,\boldsymbol{\tau}_{2k}^l + u_{2k+1}^l(t,y)\,\boldsymbol{\tau}_{2k+1}^l\right)$.

Therefore, we obtain functions $u_0^l(t,y)$ for $t \geq |y|$ from the following equation:

$$\sum_{l=0}^{2m-1} u_0^l(t,y)\,\boldsymbol{\tau}_0^l = \sum_{l=0}^{2m-1} u_0^l(t,y)\,\mathbf{e}^l$$

$$= \frac{1}{2\pi}\int_{-\pi}^{\pi} e^{\mathbf{e}(t\cos s + yi\sin s)}ds = J_0\left(\mathbf{e}i\sqrt{y^2-t^2}\right) = I_0\left(\mathbf{e}\Lambda_y\right),$$

where J_k (respectively, I_k) is the Bessel (respectively, modified Bessel) function of the first kind and kth order (Gradshteyn and Ryzhik 1980).

From equations [5.18] and [5.19], we can obtain the following Cauchy–Riemann type of conditions:

$$\frac{\partial}{\partial t}u_0^{l\oplus 1} = \frac{1}{2}u_2^l,\quad \frac{\partial}{\partial t}u_1^{l\oplus 1} = \frac{1}{2}u_3^l,\quad \frac{\partial}{\partial t}u_2^{l\oplus 1} = u_0^l + \frac{1}{2}u_4^l,$$

$$\frac{\partial}{\partial t}u_{2k-1}^{l\oplus 1} = \frac{1}{2}\left(u_{2k-3}^l + u_{2k+1}^l\right),\quad \frac{\partial u_{2k}^{l\oplus 1}}{\partial t} = \frac{1}{2}\left(u_{2(k-1)}^l + u_{2(k+1)}^l\right),\qquad [5.21]$$

for $k = 2, 3, \ldots$, and

$$\frac{\partial}{\partial y}u_0^{l\oplus 1} = -\frac{1}{2}u_1^l,\quad \frac{\partial u_1^{l\oplus 1}}{\partial y} = u_0^l - \frac{1}{2}u_4^l,\quad \frac{\partial}{\partial y}u_2^{l\oplus 1} = -\frac{1}{2}u_3^l,$$

$$\frac{\partial u_{2k+1}^{l\oplus 1}}{\partial y} = \frac{1}{2}\left(u_{2k}^l - u_{2(k+2)}^l\right),\quad \frac{\partial u_{2k+2}^{l\oplus 1}}{\partial y} = \frac{1}{2}\left(u_{2k-1}^l - u_{2k+3}^l\right),\qquad [5.22]$$

for $k = 1, 2, \ldots$. By using equations [5.21] and [5.22] and functions $u_0^l(t,y)$, we obtain, by iterative means, the function $u_k^l(t,y)$ for any $k \geq 1$, which will be used as a solution of equation [5.13]. For mathematical convenience, let us define $\sqrt{t^2-x^2} = \Lambda_x$ and $\sqrt{t^2-y^2} = \Lambda_y$ in the rest of the chapter.

THEOREM 5.3.– The pdf $f(t,x)$ of the particle position at time t for $m = 2$, $v = 1$, $\lambda = 1$ is of the following form:

$$f(t,x) = -\frac{J_0(\Lambda_x)}{2}e^{-t} + \frac{(t+x)e^{-t}}{2\Lambda_x}I_1(\Lambda_x)$$

$$-\frac{x^2e^{-t}}{2\sqrt{(t^2-x^2)^3}}\left(I_1(\Lambda_x) + J_1(\Lambda_x)\right) + \frac{t^2e^{-t}}{4(t^2-x^2)}\left(I_0(\Lambda_x)\right.$$

$$\left.+I_2(\Lambda_x) + J_0(\Lambda_x) - J_2(\Lambda_x)\right) + \delta(t-x)e^{-t} + t\delta(t-x)e^{-t}.$$

PROOF.– It follows from lemma 5.1 that, for this case, the singular part is given by

$$f_s(t,x) = e^{-t}(1+t)\delta(t-x)$$

and hence,

$$g_s(t,y) = (1+t)\delta(t-y).$$

The algebra A_0 is of the following form:

$$A_0 = \{a + e_1 b + e_2 c + e_3 d \,|\, a, b, c, d \in \mathbb{R}\},$$

where the basis $e_l = \mathbf{e}^l$, $l = 0, 1, 2, 3$, and $\mathbf{e}$ has the following matrix representation:

$$\mathbf{e} \to P = \begin{bmatrix} 0 & 1 & 0 & 0 \\ 0 & 0 & 1 & 0 \\ 0 & 0 & 1 & 0 \\ 0 & 1 & 0 & 0 \end{bmatrix}.$$

Therefore, we have $\boldsymbol{\tau}_0^0 = 1$, $\boldsymbol{\tau}_{2k}^0 = \cos ks$, $\boldsymbol{\tau}_{2k}^l = \mathbf{e}^l \cos ks$, $\boldsymbol{\tau}_{2k+1}^l = \mathbf{e}^l i \sin(k+1)s$, $l = 0, 1, 2, 3$, $k = 0, 1, 2, \ldots$.

Taking into account that $\mathbf{g}_c(\mathbf{e}(t\cos s + yi\sin s)) = e^{\mathbf{e}(t\cos s + yi\sin s)}$, we have

$$u_0^0(t,y) + \mathbf{e}u_0^1(t,y) + \mathbf{e}^2 u_0^2(t,y) + \mathbf{e}^3 u_0^3(t,y)$$

$$= \frac{1}{2\pi}\int_{-\pi}^{\pi} e^{\mathbf{e}(t\cos s + yi\sin s)}ds = I_0(\mathbf{e}\Lambda_y).$$

It is easily seen that

$$I_0(\mathbf{e}\Lambda_y) = \frac{I_0(\Lambda_y) + I_0(i\Lambda_y)}{2} + \mathbf{e}^2\left(\frac{I_0(\Lambda_y) - I_0(i\Lambda_y)}{2}\right)$$

$$= \frac{I_0(\Lambda_y) + J_0(\Lambda_y)}{2} + \mathbf{e}^2\frac{I_0(\Lambda_y) - J_0(\Lambda_y)}{2}.$$

Thus, for $t \geq |y|$ we have $u_0^1(t,y) = u_0^3(t,y) = 0$ and

$$u_0^0(t,y) = \frac{I_0(\Lambda_y) + J_0(\Lambda_y)}{2},$$

$$u_0^2(t,y) = \frac{I_0(\Lambda_y) - J_0(\Lambda_y)}{2}. \qquad [5.23]$$

It follows from the first two equations of [5.22] that

$$u_1^1 = -2\frac{\partial}{\partial y}u_0^2 = -\frac{\partial\left[I_0(\Lambda_y) - J_0(\Lambda_y)\right]}{\partial y} = \frac{y}{\Lambda_y}(I_1(\Lambda_y) + J_1(\Lambda_y)),$$

$$u_1^3 = -2\frac{\partial}{\partial y}u_0^0 = -\frac{\partial\left[I_0(\Lambda_y) + J_0(\Lambda_y)\right]}{\partial y} = \frac{y}{\Lambda_y}(I_1(\Lambda_y) - J_1(\Lambda_y)),$$

$$u_1^0 = -2\frac{\partial}{\partial y}u_0^1 = 0, \quad u_1^2 = -2\frac{\partial}{\partial y}u_0^3 = 0.$$

Then, from the Cauchy–Riemann type of conditions [5.21], it follows that

$$u_2^0(t,y) = 2\frac{\partial u_0^1(t,y)}{\partial t} = 0;$$

$$u_2^1(t,y) = 2\frac{\partial u_0^2(t,y)}{\partial t} = \frac{\partial\left(I_0(\Lambda_y) - J_0(\Lambda_y)\right)}{\partial t} = \frac{t}{\Lambda_y}(I_1(\Lambda_y) + J_1(\Lambda_y));$$

$$u_2^2(t,y) = 2\frac{\partial u_0^3(t,y)}{\partial y} = 0;$$

$$u_2^3(t,y) = 2\frac{\partial u_0^0(t,y)}{\partial t} = \frac{\partial\left[I_0(\Lambda_y) + J_0(\Lambda_y)\right]}{\partial t} = \frac{t}{\Lambda_y}(I_1(\Lambda_y) - J_1(\Lambda_y)).$$

Similarly, for u_3^l we have

$$u_3^0 = 2\frac{\partial}{\partial t}u_1^1 = 2\frac{\partial}{\partial t}\left[\frac{y}{\Lambda_y}\left(I_1(\Lambda_y) + J_1(\Lambda_y)\right)\right]$$

$$= -\frac{2ty}{\sqrt{(t^2-y^2)^3}}\left(I_1(\Lambda_y) + J_1(\Lambda_y)\right)$$

$$+\frac{ty}{t^2-y^2}\left(I_0(\Lambda_y) + I_2(\Lambda_y) + J_0(\Lambda_y) - J_2(\Lambda_y)\right);$$

$$u_3^2 = 2\frac{\partial}{\partial t}u_1^3 = 2\frac{\partial}{\partial t}\left[\frac{y}{\Lambda_y}\left(I_1(\Lambda_y) - J_1(\Lambda_y)\right)\right]$$

$$= -\frac{2ty}{\sqrt{(t^2-y^2)^3}}\left(I_1(\Lambda_y) - J_1(\Lambda_y)\right)$$

$$+\frac{2ty}{t^2-y^2}(I_0(\Lambda_y) + I_2(\Lambda_y) - J_0(\Lambda_y) + J_2(\Lambda_y)).$$

It is easily seen that $u_3^1 = u_3^3 = 0$.

From [5.21], it follows that

$$u_4^0 = 2\frac{\partial u_2^1}{\partial t} - 2u_0^0 = 2\frac{\partial}{\partial t}\frac{t}{\Lambda_y}\left(I_1(\Lambda_y) + J_1(\Lambda_y)\right) - 2u_0^0$$

$$= \frac{-2y^2}{\sqrt{(t^2-y^2)^3}}\left(I_1(\Lambda_y) + J_1(\Lambda_y)\right)$$

$$+\frac{t^2}{t^2-y^2}\left(I_0(\Lambda_y) + I_2(\Lambda_y) + J_0(\Lambda_y) - J_2(\Lambda_y)\right) - I_0(\Lambda_y) - J_0(\Lambda_y);$$

$$u_4^2 = 2\left(\frac{\partial u_2^3}{\partial t} - u_0^2\right)$$
$$= \frac{-2y^2}{\sqrt{(t^2-y^2)^3}}\left(I_1(\Lambda_y) - J_1(\Lambda_y)\right) + \frac{t^2}{t^2-y^2}\left(I_0(\Lambda_y)\right.$$
$$\left. + I_2(\Lambda_y) - J_0(\Lambda_y) + J_2(\Lambda_y)\right) - I_0(\Lambda_y) + J_0(\Lambda_y).$$

It is also easily verified that $u_4^1 = u_4^3 = 0$.

By using well-known integral expressions for Bessel functions (Bateman 1954; Gradshteyn and Ryzhik 1980), we have

$$\int_{-t}^{t} u_0^0 dy = \sinh t + \sin t, \qquad \int_{-t}^{t} u_0^2 dy = \sinh t - \sin t,$$
$$\int_{-t}^{t} u_1^1 dy = \int_{-t}^{t} u_1^3 dy = 0,$$
$$\int_{-t}^{t} u_2^1 dy = 2\int_{-t}^{t} \frac{\partial u_0^2}{\partial t} dy = 2\left(\frac{\partial}{\partial t}\int_{-t}^{t} u_0^2 dy - u_0^2(t,t) - u_0^2(t,-t)\right)$$
$$= 2\cosh t - 2\cos t,$$
$$\int_{-t}^{t} u_2^3 dy = 2\int_{-t}^{t} \frac{\partial u_0^0}{\partial t} dy = 2\left(\frac{\partial}{\partial t}\int_{-t}^{t} u_0^0 dy - u_0^0(t,t) - u_0^0(t,-t)\right)$$
$$= 2\cosh t + 2\cos t - 4.$$

As an example, we obtain the pdf for the case when $m = 1$. For this case, $e_1 = 1$, and hence, we can consider the functions $\sum_{l=0}^{4} u_k^l(t,y)$, $k = 0, 1, 2, \ldots$ as solutions of [5.13] for $m = 1$.

Let us consider the function $g(t,y) = g_c(t,y) + g_s(t,y)$, where $t \leq |y|$,

$$g_c(t,y)$$
$$= \frac{1}{2}\left(u_0^0(t,y) + u_0^2(t,y)\right) + \frac{1}{4}\left(u_1^1(t,y) + u_1^3(t,y) + u_2^1(t,y) + u_2^3(t,y)\right)$$
$$= \frac{I_0(\Lambda_y)}{2} + \frac{t+y}{2\Lambda_y} I_1(\Lambda_y)$$

and $g_s(t,y) = \delta(t-y)$.

It is easily seen that the function $g_c(t,y)$ is a solution of the following equation for $t < y$:

$$\left(\frac{\partial^2}{\partial t^2} - \frac{\partial^2}{\partial y^2}\right) g(t,y) - g(t,y) = 0, \qquad [5.24]$$

In addition, we have $\lim_{y\uparrow t} g_c(t,y) = \frac{1}{2}(1+t)$ and $\lim_{y\downarrow -t} g_c(t,y) = \frac{1}{2}$.

Thus, $f(t,x) = e^{-t}g(t,x)$ is a solution of the equation:

$$\left(\frac{\partial}{\partial t} - \frac{\partial}{\partial x} + 1\right)\left(\frac{\partial}{\partial t} + \frac{\partial}{\partial x} + 1\right) f_c(t,x) - f_c(t,x) = 0, \tag{5.25}$$

$$f_s(t,x) = \delta(t-x)e^{-t}.$$

In addition, $f_c(t,x)$ satisfies the following conditions:

$$\lim_{x\uparrow t} f_c(t,x) = \frac{1}{2}\left(e^{-t} + te^{-t}\right), \quad \lim_{x\downarrow -t} f_c(t,x) = \frac{1}{2}e^{-t},$$

and for all $t > 0$ we have $\int_{-t}^{t} f(t,x)dx = 1$.

For a small $\varepsilon > 0$, consider the probability $P\{0 < t - x(t) < \varepsilon\}$.

Let us verify that

$$\lim_{x\uparrow t} f_c(t,x) = \lim_{\varepsilon\downarrow 0} \frac{P\{0 < t - x(t) < \varepsilon\}}{\varepsilon}$$

or equivalently,

$$\lim_{\varepsilon\downarrow 0} \frac{P\{0 < t - x(t) < \varepsilon\}}{\varepsilon} = \frac{1}{2}\left(e^{-t} + te^{-t}\right).$$

Indeed, it is easily seen that

$$P\{0 < t - x(t) \le \varepsilon\} = P\left\{t - \frac{\varepsilon}{2} \le \theta_1 < t\right\}$$

$$+ \int_0^t P\left\{\theta_3 \ge t - u, \theta_2 \le \frac{\varepsilon}{2}, \theta_1 \in du\right\} + o(\varepsilon),$$

where θ_i, $i = 1,2,3$ are independent exponentially distributed random variables. The random variable θ_1 represents the time of the first velocity alternation, θ_2 is the time between the first and the second velocity alternations, and θ_3 is the time between the second and the third velocity alternations.

We have that $P\left\{t - \frac{\varepsilon}{2} \le \theta_1 < t\right\} = e^{-t+\frac{\varepsilon}{2}} - e^{-t}$. It is then easy to calculate

$$\int_0^t P\left\{\theta_3 \ge t - u,\ \theta_2 \le \frac{\varepsilon}{2}, \theta_1 \in du\right\} = \left(1 - e^{-\frac{\varepsilon}{2}}\right)\int_0^t e^{-t+u}e^{-u}du$$

$$= \left(1 - e^{-\frac{\varepsilon}{2}}\right)te^{-t}.$$

Hence, it is evident that

$$\lim_{\varepsilon \downarrow 0} \frac{P\{0 < t - x(t) < \varepsilon\}}{\varepsilon} = \frac{1}{2}\left(e^{-t} + te^{-t}\right).$$

Similarly,

$$P\{t + x(t) \leq \varepsilon\} = P\left\{\theta_2 \geq t - \frac{\varepsilon}{2}, \, \theta_1 \leq \frac{\varepsilon}{2}\right\} + o(\varepsilon).$$

This implies that

$$\lim_{\varepsilon \downarrow 0} \frac{P\{t + x(t) < \varepsilon\}}{\varepsilon} = \frac{1}{2} e^{-t} = \lim_{x \downarrow -t} f_c(t, x).$$

Therefore, $f_c(t, x)$ is a solution of the Goursat problem for the linear second-order hyperbolic equation that ensures the uniqueness of the solution of equation [5.25]. It means that $f(t, x)$ is the pdf of the particle position at time t for $m = 1$.

It is relevant to remark that $f(t, x)$ coincides with the result obtained in Pinsky (1991).

We now turn to the case $m = 2$ and continue to calculate integrals of u_k^l.

From $u_4^0 = 2\frac{\partial u_2^1}{\partial t} - 2u_0^0$, it follows that

$$\int_{-t}^{t} u_4^0 dy = 2\left(\frac{\partial}{\partial t}\int_{-t}^{t} u_2^1 dy - u_2^1(t, t) - u_2^1(t, -t)\right) - 2\sinh t - 2\sin t$$

$$= 4(\sinh t + \sin t - t) - 2\sinh t - 2\sin t = 2\sinh t + 2\sin t - 4t.$$

Next, from $u_4^2 = 2\frac{\partial u_2^3}{\partial t} - 2u_0^2$, it follows that

$$\int_{-t}^{t} u_4^2 dy = 2\left(\frac{\partial}{\partial t}\int_{-t}^{t} u_2^3 dy - u_2^3(t, t) - u_2^3(t, -t)\right) - 2\sinh t + 2\sin t$$

$$= 4\sinh t - 4\sin t - 2\sinh t + 2\sin t = 2\sinh t - 2\sin t.$$

For $t \leq |y|$, we introduce the function $g(t, y) = g_c(t, y) + g_s(t, y)$, where

$$u_c(t, y)$$

$$= \frac{1}{2}u_0^2(t, y) + \frac{1}{4}\left(u_1^1(t, y) + u_1^3(t, y) + u_2^1(t, y) + u_2^3(t, y) + u_4^0(t, y)\right),$$

$$g_s(t, y) = \delta(t - y) + t\delta(t - y). \qquad [5.26]$$

By construction, the function $g_c(t,y)$ is a solution of the following equation:

$$\left(\frac{\partial^2}{\partial t^2} - \frac{\partial^2}{\partial y^2}\right)^2 g(t,y) - g(t,y) = 0. \qquad [5.27]$$

Therefore, the function $f_c(t,x) = e^{-t} g_c(t,x)$ is a solution of equation [5.10] for $m = 2$ ($\lambda = 1, v = 1$).

We establish that $f(t,x) = f_c(t,x) + e^{-t} g_s(t,x)$. Taking into account the values of integrals of functions that are involved in the expression for $g_c(t,y)$, we have that $\int_{-t}^{t} f(t,x)dx = 1$, for all $t \geq 0$.

Let us show that

$$\lim_{x \uparrow t} f_c(t,x) = \lim_{\varepsilon \downarrow 0} \frac{P\{0 < t - x(t) < \varepsilon\}}{\varepsilon}$$

and

$$\lim_{x \downarrow -t} f_c(t,x) = \lim_{\varepsilon \downarrow 0} \frac{P\{t + x(t) < \varepsilon\}}{\varepsilon}.$$

From lemma 5.2, it follows that for $m = 2$ we have

$$\lim_{\varepsilon \downarrow 0} \frac{P\{0 < t - x(t) < \varepsilon\}}{\varepsilon} = \frac{1}{2} t e^{-t}. \qquad [5.28]$$

and

$$\lim_{\varepsilon \downarrow 0} \frac{P\{t + x(t) < \varepsilon\}}{\varepsilon} = 0.$$

It can be easily verified that $\lim_{y \uparrow t} u_4^0(t,y) = 0$, $\lim_{y \uparrow t} u_0^2(t,y) = 0$, and consequently

$$\lim_{y \uparrow t} g_c(t,y) = \lim_{y \uparrow t} \frac{t+y}{2\Lambda_y} I_1(\Lambda_y) = \frac{t}{2},$$

$$\lim_{y \downarrow -t} g_c(t,y) = \lim_{y \downarrow -t} \frac{t+y}{2\Lambda_y} I_1(\Lambda_y) = 0. \qquad [5.29]$$

Thus,

$$\lim_{x \uparrow t} f_c(t,x) = \frac{1}{2} t e^{-t} = \lim_{\varepsilon \downarrow 0} \frac{P\{t - x(t) < \varepsilon\}}{\varepsilon},$$

$$\lim_{x \downarrow -t} f_c(t,x) = 0 = \lim_{\varepsilon \downarrow 0} \frac{P\{t + x(t) < \varepsilon\}}{\varepsilon}. \qquad [5.30]$$

Let us show that conditions [5.29], in addition to condition

$$\int_{-t}^{t} g\left(t,y\right)e^{-t}dx=1,$$

ensure the uniqueness of the solution $g_c\left(t,y\right)$ for equation [5.27], and, consequently, the uniqueness of the solution $f_c\left(t,x\right)$ of equation [5.21].

It is not hard to verify that each solution of equation [5.25] is a solution of equation [5.24]. By changing the variables $s=t+y, p=t-y$, we reduce equation [5.24] to

$$\frac{\partial^4}{\partial s^2\partial p^2}G\left(s,p\right)-G\left(s,p\right)=0. \qquad [5.31]$$

Let us take the Fourier transform $\hat{G}\left(s,\alpha\right)=\int_{-\infty}^{\infty}G\left(s,p\right)e^{i\alpha p}dp$ in equation [5.31], thus we obtain the ordinary differential equation of order 4. Now, by considering that $\lim_{y\downarrow -t}g_c\left(t,y\right)=0$, we have

$$\hat{G}\left(0,\alpha\right)=0. \qquad [5.32]$$

Hence, at most four independent solutions of the ordinary differential equation [5.31] satisfy the initial condition [5.32] for each α. By taking the inverse Fourier transform, we obtain four independent solutions of equation [5.27] with the condition $\lim_{x\downarrow -t}g_c\left(t,x\right)=0$, and just two of them satisfy equation [5.27], but not equation [5.24]. By construction, one of these solutions $g_c\left(t,y\right)$ is given by equation [5.26]. As another solution, we can take

$$g_2(t,y)=u_0^2\left(t,y\right)+u_4^0\left(t,y\right).$$

We can easily verify that no linear combination $c(t,y)$ of functions $g_c\left(t,y\right)$ and $g_2(t,x)$ satisfies conditions [5.30] and

$$\int_{-t}^{t}\left(c\left(t,x\right)+g_s\left(t,y\right)\right)e^{-t}dx=1$$

for all $t>0$ except the solution $g_c\left(t,y\right)$.

Similarly, with the pdf $f\left(t,x\right)$ of the particle position for $m=2$ we can also obtain solutions of equation [5.10] with conditions [5.8] and [5.12] for each $m>2$.

5.2. Distribution of limiting position of fading evolution

This section deals with the distribution of the following random series:

$$\sum_{n=1}^{\infty}a^n\theta_n,$$

where $0 < a < 1$, and θ_n, $n = 1, 2, \ldots$ are i.i.d. random variables with uniform or Erlang distributions.

Based on this result, we obtain the distribution of the limiting position of a particle whose trajectory is determined by the fading random evolution in Erlang or uniform media on the line or in multidimensional space.

5.2.1. *Distribution of random power series in cases of uniform and Erlang distributions*

We calculate the distribution of a random power series in the cases of the uniform and 2-Erlang distributions of terms, and by using this distribution we obtain the distribution of a fading evolution.

Let us consider a random power series $\nu = \sum_{n=1}^{\infty} a^{2n-2}\theta_{2n-1}$, where $0 < a < 1$, and θ_n are i.i.d. random variables. Let θ_1 be uniformly distributed on the interval $[0, 1]$.

Denote by $F_\nu(x)$ the distribution function of the sum of the power series ν.

LEMMA 5.4.– The function $F_\nu(x)$ is given by

for $0 \le x \le 1$,

$$F_\nu(x) = c_0 \left(\sum_{n=0}^{+\infty} (-1)^n a^{n(n+1)} e^{-\frac{x}{a^{2n}}} \right.$$

$$\left. + \sum_{m=1}^{+\infty} (-1)^{m+1} a^{m(m+1)} e^{-xa^{2m}} \right), \qquad [5.33]$$

and for $1 < x \le \frac{1}{1-a^2}$,

$$F_\nu(x) = c_0 \left(\sum_{n=0}^{+\infty} (-1)^{n-1} a^{n(n+1)} \left(e^{\frac{1-x}{a^{2n}}} - e^{-\frac{x}{a^{2n}}} \right) \right.$$

$$\left. + \sum_{m=1}^{+\infty} (-1)^m a^{m(m+1)} \left(e^{(1-x)a^{2m}} - e^{-xa^{2m}} \right) \right), \qquad [5.34]$$

where

$$c_0^{-1} = \sum_{n=0}^{+\infty} (-1)^{n-1} a^{n(n+1)} \left(e^{\frac{a^2}{a^{2n}(a^2-1)}} - e^{\frac{1}{a^{2n}(a^2-1)}} \right)$$

$$+ \sum_{m=1}^{+\infty} (-1)^m a^{m(m+1)} \left(e^{\frac{a^{2m+2}}{a^2-1}} - e^{\frac{a^{2m}}{a^2-1}} \right).$$

PROOF.– For $F_\nu(x)$, we have

$$F_\nu(x) = P\left\{\theta_1 + a^2\nu' \le x\right\} = \int_0^x P\left\{u + a^2\nu' \le x\right\} du$$

$$= \int_0^x P\left\{\eta' \le \frac{x-u}{a^2}\right\} du.$$

where η' is a random variable with the same distribution as η.

Therefore, we have

$$F_\nu(x) = \int_0^x F_\nu\left\{\frac{x-u}{a^2}\right\} du, \quad 0 \le x \le 1,$$

$$F_\nu(x) = \int_0^1 F_\nu\left\{\frac{x-u}{a^2}\right\} du, \; x > 1. \qquad [5.35]$$

We will find a solution of equation [5.35] in the following form:

$$F_\nu(x) = \sum_{n=0}^{+\infty} c_n e^{-\frac{x}{a^{2n}}} + \sum_{m=1}^{+\infty} b_m e^{-xa^{2m}}. \qquad [5.36]$$

Substituting [5.36] into [5.35], we have for $0 \le x \le 1$

$$\sum_{n=0}^{+\infty} c_n e^{-\frac{x}{a^{2n}}} + \sum_{m=1}^{+\infty} b_m e^{-xa^{2m}}$$

$$= \sum_{n=0}^{+\infty} c_n \int_0^x e^{-\frac{x-u}{a^{2n+2}}} du + \sum_{m=1}^{+\infty} b_m \int_0^x e^{-(x-u)a^{2m-2}} du$$

$$= \sum_{n=0}^{+\infty} c_n a^{2n+2}\left(1 - e^{-\frac{x}{a^{2n+2}}}\right) + \sum_{m=1}^{+\infty} \frac{b_m}{a^{2m-2}}\left(1 - e^{-xa^{2m-2}}\right).$$

After equating coefficients of like exponential terms, we have

$$c_n a^{2n+2} = -c_{n+1}, \quad n = 0, 1, 2, \ldots,$$

$$b_1 = -c_0, \quad b_m = -a^{2m-2} b_{m-1}, \quad m = 2, 3, \ldots.$$

Therefore, $c_n = (-1)^n a^{n(n+1)} c_0$ and $b_{m+1} = (-1)^{m+1} a^{m(m+1)} c_0$.

It is easily verified that

$$\sum_{n=0}^{+\infty} c_n a^{2n+2} + \sum_{m=1}^{+\infty} \frac{b_m}{a^{2m-2}} = 0. \qquad [5.37]$$

Since θ_n is an absolutely continuous random variable, then the function $F_\nu(x)$ is absolutely continuous, i.e. there exists the pdf $f_\nu(x) = \frac{d}{dx} F_\nu(x)$. It is easily seen that the support of $f_\nu(x)$ is the segment $\left[0, \frac{1}{1-a^2}\right]$, and, consequently, $F_\nu\left(\frac{1}{1-a^2}\right) = 1$.

Thus, we have

$$F_\nu(x) = c_0 \left(\sum_{n=0}^{+\infty} (-1)^n a^{n(n+1)} e^{-\frac{x}{a^{2n}}} + \sum_{m=1}^{+\infty} (-1)^{m+1} a^{m(m+1)} e^{-xa^{2m}} \right),$$

for $0 \leq x \leq 1$.

It is easily verified that $F_\nu(0) = 0$. From [5.36], it follows that

$$F_\nu(x) = c_0 \sum_{n=0}^{+\infty} (-1)^{n-1} a^{n(n+1)} \left(e^{\frac{1-x}{a^{2n}}} - e^{-\frac{x}{a^{2n}}} \right)$$

$$+ c_0 \sum_{m=1}^{+\infty} (-1)^m a^{m(m+1)} \left(e^{(1-x)a^{2m}} - e^{-xa^{2m}} \right).$$

for $1 < x \leq \frac{1}{1-a^2}$.

Taking into account that $F_\nu\left(\frac{1}{1-a^2}\right) = 1$, we obtain

$$c_0 = \left(\sum_{n=0}^{+\infty} (-1)^{n-1} a^{n(n+1)} \left(e^{\frac{a^2}{a^{2n}(a^2-1)}} - e^{\frac{1}{a^{2n}(a^2-1)}} \right) \right.$$

$$\left. + \sum_{m=1}^{+\infty} (-1)^m a^{m(m+1)} \left(e^{\frac{a^{2m+2}}{a^2-1}} - e^{\frac{a^{2m}}{a^2-1}} \right) \right)^{-1}.$$

Next, it follows from [5.35] that the function $F_\nu(x)$ is monotonic. Indeed, as a consequence of lemma 5.5, all coefficients of the expansion [5.36] are bounded. Therefore, $F_\nu(x)$ is continuously differentiable.

Let $\frac{d}{dx} F_\nu(x) = f_\nu(x)$, and using equations [5.35], we have the pdf

$$f_\nu(x) = \frac{1}{a^2} \int_0^x f_\nu\left(\frac{x-u}{a^2}\right) du = \frac{1}{a^2} \int_0^x (x-u)\, e^{-(x-u)} f_\nu\left(\frac{u}{a^2}\right) du. \quad [5.38]$$

Since $f_\nu(x)$ is continuous, then it follows from [5.38] that $f_\nu(x)$ is continuously differentiable, i.e. there exists $f_\nu'(x)$, $x \geq 0$. Suppose that for a segment $[0, \alpha_0]$, $\alpha_0 > 0$ the function $f_\nu(x) \geq 0$, i.e. $F_\nu(x)$ is a non-decreasing function on $[0, \alpha_0]$. It is easily verified that there exists $\alpha_1 > 0$ such that $0 < \alpha_1 < \alpha_0$ and $f_\nu'(x) \geq 0$ for each $x \in [0, \alpha_1]$. By continuing this procedure, we obtain that for any $n \in \mathbb{N}$, there exists an $\alpha_n > 0$ such that $f_\nu^{(n)}(x) \geq 0$ for all $x \in [0, \alpha_n]$.

Hence, all derivatives $f_\nu^{(n)}(0)$, $n \geq 1$, of $F_\eta(x)$ are non-negative and consequently

$$f_\nu^{'}(x) = \sum_{n=1}^{\infty} \frac{f_\nu^{(n)}(0)}{(n-1)!} x^{n-1} \geq 0, \quad x \geq 0.$$

Therefore, $F_\eta(x)$ is a non-decreasing function for all $x \geq 0$.

The case where $F_\nu(x)$ is non-increasing on an initial segment $[0, \alpha_0]$ can be studied in a similar manner.

Thus, the solution $F_\nu(x)$ of equation [5.35] is a monotonic function. From [5.34] and [5.36], it follows that $\lim_{x \to +\infty} F_\nu(x) = 1$.

Hence, $F_\nu(x)$ is a distribution function. Let us show that there is not another distribution function satisfying equation [5.35]. Indeed, suppose a distribution function $\tilde{F}_\nu(x)$ that satisfies [5.35] and $\tilde{F}_\nu(x) \neq F_\nu(x)$, then the function $G(x) = F_\nu(x) - \tilde{F}_\nu(x)$ is a solution of equation [5.35], but this is impossible as $G(x)$ is not a monotonic function, except for the case where $G(x) = 0$.

Now consider the random power series $\eta = \sum_{n=1}^{\infty} a^{2n-2}\theta_{2n-1}$, where $0 < a < 1$, and θ_n are i.i.d. random variables with the Erlang-2 distribution of the following form:

$$f_k(t) = \frac{dF_k(t)}{dt} = \lambda^2 t e^{-\lambda t} \mathbb{I}_{\{t \geq 0\}}, \quad \lambda > 0.$$

Denote by $F_\eta(x) = P\{\eta \leq x\}$ the cdf of η. To simplify expressions, we consider $\lambda = 1$.

LEMMA 5.5.– The function $F_\eta(x)$ has the form

$$F_\eta(x) = 1 + (a_{01}x + a_{02})e^{-x} + (a_{11}x + a_{12})e^{-\frac{x}{a^2}} + \ldots$$
$$+ (a_{n1}x + a_{n2})e^{-\frac{x}{a^{2n}}} + \ldots,$$

where for $n= 1, 2, \ldots$

$$a_{n1} = a_{n-11} \frac{a^{4n-2}}{(a^{2n}-1)^2};$$

$$a_{n2} = -a_{n-11} \frac{2a^{6n-2}}{(a^{2n}-1)^3} + a_{n-12} \frac{a^{4n}}{(a^{2n}-1)^2};$$

and

$$a_{01} = \frac{S_2 + S_3}{S_1 S_3 + S_2 S_4} \qquad a_{02} = \frac{S_4 - S_1}{S_1 S_3 + S_2 S_4},$$

where

$$S_1 = \left(1 + \frac{a^2}{(a^2-1)^2} + \frac{a^8}{(a^2-1)^2(a^4-1)^2} + \frac{a^{18}}{(a^2-1)^2(a^4-1)^2(a^6-1)^2} + \ldots\right)$$
$$+2\left(\frac{a^8}{(a^2-1)^3(a^4-1)} + \frac{a^{18}}{(a^2-1)^3(a^4-1)^2(a^6-1)} + \frac{a^{18}}{(a^2-1)^2(a^4-1)^3(a^6-1)} + \ldots\right);$$

$$S_2 = \frac{a^2}{1-a^2} + \frac{a^8}{(1-a^2)^2(1-a^4)} + \frac{a^{18}}{(1-a^2)^2(1-a^4)^2(1-a^6)} + \ldots;$$

$$S_3 = \frac{2a^2}{(a^2-1)^3} + \frac{2a^{12}}{(a^2-1)^2(a^4-1)^3} + \frac{2a^{12}}{(a^2-1)^3(a^4-1)^2} + \frac{2a^{24}}{(a^2-1)^2(a^4-1)^2(a^6-1)^3} + \ldots;$$

$$S_4 = 1 + \frac{a^4}{(a^2-1)^2} + \frac{a^{12}}{(a^2-1)^2(a^4-1)^2} + \frac{a^{24}}{(a^2-1)^2(a^4-1)^2(a^6-1)^2} + \ldots;$$

PROOF.– For $F_\eta(x)$, we have

$$F_\eta(x) = P\{\theta_1 + a^2\eta' \le x\} = \int_0^x ue^{-u} P\{u + a^2\eta' \le x\}\, du$$
$$= \int_0^x ue^{-u} P\left\{\eta' \le \frac{x-u}{a^2}\right\} du.$$

where η' is a random variable with the same distribution as that of η.

Hence,

$$F_\eta(x) = \int_0^x ue^{-u} F_\eta\left\{\frac{x-u}{a^2}\right\} du. \qquad [5.39]$$

It is easy to see that $F_\eta(0) = 0$. We will find $F_\eta(x)$ in the form of the following series:

$$F_\eta(x) = 1 + (a_{01}x + a_{02})\, e^{-x} + (a_{11}x + a_{12})\, e^{-\frac{x}{a^2}} + \ldots$$
$$+ (a_{n1}x + a_{n2})\, e^{-\frac{x}{a^{2n}}} + \ldots \qquad [5.40]$$

Substituting [5.40] into [5.39], we have

$$F_\eta(x) = \int_0^x ue^{-u}\left[1 + \left(a_{01}\frac{x-u}{a^2} + a_{02}\right) e^{-\frac{x-u}{a^2}}\right.$$
$$\left. + \left(a_{11}\frac{x-u}{a^2} + a_{12}\right) e^{-\frac{x-u}{a^4}} + \left(a_{21}\frac{x-u}{a^2} + a_{22}\right) e^{-\frac{x-u}{a^6}} + \ldots\right] du$$
$$= 1 - xe^{-x} - e^{-x} + a_{01}a^2 \frac{(a^2x - x + 2a^2)\, e^{-x} + (a^2x - x - 2a^2)\, e^{-\frac{x}{a^2}}}{(a^2-1)^3}$$
$$+ a_{02}a^2 \frac{(x - a^2x - a^2)\, e^{-x} + a^2 e^{-\frac{x}{a^2}}}{(a^2-1)^2}$$
$$+ a_{11}a^6 \frac{(a^4x - x + 2a^4)\, e^{-x} + (a^4x - x + 2a^4)\, e^{-\frac{x}{a^4}}}{(a^4-1)^3}$$
$$+ a_{12}a^4 \frac{(x - a^4x - a^4)\, e^{-x} + a^4 e^{-\frac{x}{a^4}}}{(a^4-1)^2}$$
$$+ a_{21}a^{10} \frac{(a^6x - x + 2a^6)\, e^{-x} + (a^6x - x + 2a^6)\, e^{-\frac{x}{a^6}}}{(a^6-1)^3}$$
$$+ a_{22}a^6 \frac{(x - a^6x - a^6)\, e^{-x} + a^6 e^{-\frac{x}{a^6}}}{(a^6-1)^2} + \ldots. \qquad [5.41]$$

Whence, taking into account [5.40], for the term xe^{-x} to be held constant, we have:

$$a_{01} = -1 + a_{01}\frac{a^2}{(a^2-1)^2} + a_{02}\frac{a^2}{1-a^2} + a_{11}\frac{a^6}{(a^4-1)^2} + a_{12}\frac{a^4}{1-a^4} + \ldots$$

$$+a_{21}\frac{a^{10}}{(a^6-1)^2} + a_{22}\frac{a^6}{1-a^6} + \ldots + a_{n1}\frac{a^{4n+2}}{(a^{2n+2}-1)^2} + a_{n2}\frac{a^{2n+2}}{1-a^{2n+2}}$$

$$+\ldots.$$

For the term e^{-x} to be held constant:

$$a_{02} = -1 + a_{01}\frac{2a^4}{(a^2-1)^3} - a_{02}\frac{a^4}{(a^2-1)^2} + a_{11}\frac{2a^{10}}{(a^4-1)^3} - a_{12}\frac{a^8}{(a^4-1)^2}$$

$$+a_{21}\frac{2a^{16}}{(a^6-1)^3} - a_{22}\frac{a^{12}}{(a^6-1)^2} + \ldots + a_{n1}\frac{2a^{6n+4}}{(a^{2n+2}-1)^3} - a_{n2}\frac{a^{4n+4}}{(a^{2n+2}-1)^2}$$

$$+\ldots.$$

For the term $x\,e^{-\frac{x}{a^{2n}}}$ to be held constant:

$$a_{n1} = a_{n-11}\frac{a^{4n-2}}{(a^{2n-1})^2};$$

and for the term $e^{-\frac{x}{a^{2n}}}$ to be held constant:

$$a_{n2} = -a_{n-11}\frac{2a^{6n-2}}{(a^{2n-1})^3} + a_{n-12}\frac{a^{4n}}{(a^{2n-1})^2}; \quad n = 1, 2, \ldots. \qquad [5.42]$$

This implies the following relations:

$$1 = a_{01}\left[\left(1 + \frac{a^2}{(a^2-1)^2} + \frac{a^8}{(a^2-1)^2(a^4-1)^2}\right.\right.$$

$$\left. + \frac{a^{18}}{(a^2-1)^2(a^4-1)^2(a^6-1)^2} + \ldots\right)$$

$$+2\left(\frac{a^8}{(a^2-1)^3(a^4-1)} + \frac{a^{18}}{(a^2-1)^3(a^4-1)^2(a^6-1)}\right.$$

$$\left.\left. + \frac{a^{18}}{(a^2-1)^2(a^4-1)^3(a^6-1)} + \ldots\right)\right]$$

$$+a_{02}\left(\frac{a^2}{1-a^2}+\frac{a^8}{(1-a^2)^2(1-a^4)}\right.$$

$$\left.+\frac{a^{18}}{(1-a^2)^2(1-a^4)^2(1-a^6)}+\dots\right); \qquad [5.43]$$

and

$$1=a_{01}\left(\frac{2a^2}{(a^2-1)^3}+\frac{2a^{12}}{(a^2-1)^2(a^4-1)^3}+\frac{2a^{12}}{(a^2-1)^3(a^4-1)^2}\right.$$

$$+\frac{2a^{24}}{(a^2-1)^2(a^4-1)^2(a^6-1)^3}$$

$$\left.+\frac{2a^{24}}{(a^2-1)^2(a^4-1)^3(a^6-1)^2}+\frac{2a^{24}}{(a^2-1)^3(a^4-1)^2(a^6-1)^2}+\dots\right)$$

$$-a_{02}\left(1+\frac{a^4}{(a^2-1)^2}+\frac{a^{12}}{(a^2-1)^2(a^4-1)^2}\right.$$

$$\left.+\frac{a^{24}}{(a^2-1)^2(a^4-1)^2(a^6-1)^2}+\dots\right). \qquad [5.44]$$

By using the D'Alambert test for convergence, it is easy to check that all series involved in equations [5.43] and [5.44] are convergent.

We should note that coefficient a_{01} in [5.43] is related to the series

$$1+\frac{a^2}{(a^2-1)^2}+\frac{a^8}{(a^2-1)^2(a^4-1)^2}+\frac{a^{18}}{(a^2-1)^2(a^4-1)^2(a^6-1)^2}+\dots$$

$$=\prod_{k=1}^{\infty}\left(1-a^{2k}\right)^{-1}.$$

By solving equations [5.43] and [5.44], we have

$$a_{01}=\frac{S_2+S_3}{S_1S_3+S_2S_4};$$

$$a_{02}=\frac{S_4-S_1}{S_1S_3+S_2S_4}.$$

Whence, by using [5.42], we calculate all other coefficients of expansion [5.40].

From [5.42], it also follows that $F_\eta(0) = 1 + a_{02} + a_{12} + a_{22} + \ldots = 0$. Next, it follows from [5.39] that the function $F_\eta(x)$ is monotonic. The proof is much the same as for the case of the uniform distribution.

From [5.40] and [5.42], it follows that $\lim_{x\to+\infty} F_\eta(x) = 1$.

Therefore, $F_\eta(x)$ is a unique distribution function satisfying equation [5.39] and is the distribution of η.

REMARK 5.7.– *It is easily verified that in the case of an arbitrary parameter $\lambda > 0$ of Erlang-2, the distribution function $F_\eta(x)$ has the following form:*

$$F_\eta(x) = 1 + (a_{01}x + a_{02})\,e^{-\lambda x} + (a_{11}x + a_{12})\,e^{-\frac{\lambda x}{a^2}} + \ldots$$
$$+ (a_{n1}x + a_{n2})\,e^{-\frac{\lambda x}{a^{2n}}} + \ldots.$$

5.2.2. *The distribution of the limiting position*

Let $\{\xi(t)\}$ be a renewal process, which is defined by $\xi(t) = \max\{n \geq 0 : \tau_n \leq t\}$, $t>0$, where $\tau_n = \sum_{k=0}^{n} \theta_k$ and $\theta_k > 0$, $k = 0, 1, 2, \ldots$ are i.i.d. random variables.

The process $\{x(t)\}$, which is defined by the following equation for $0 < a < 1$,

$$x(t) = v \int_0^t (-a)^{\xi(s)} ds,$$

is called a fading random evolution with the switching process $\xi(t)$ (or in semi-Markov media $\xi(t)$).

A fading evolution describes the random movement of a particle such that after each switching direction of the particle movement its absolute velocity decreases and, therefore, the particle loses energy with time. Thus, a fading evolution may be a model of one-dimensional movement, where a particle loses its energy under the action of friction or loses energy colliding with other particles.

Over time, the particle stops its movement and our aim is to obtain the distribution of a stopping point for the particle, i.e. the distribution of the following random variable:

$$\sigma = v \int_0^\infty (-a)^{\xi(s)} ds.$$

Since θ_k, $k = 1, 2, \ldots$ are identically distributed and

$$\sigma = (\theta_1 + a^2\theta_3 + \ldots) - a(\theta_2 + a^2\theta_4 + \ldots),$$

the cdf $F_{\sigma}(x) = P\{\sigma \leq x\}$, which is the distribution of the limiting position of the fading evolution $x(t)$, and can be represented as follows:

$$F_{\sigma}(x) = P\{\sigma \leq x\} = P\left\{\left(\theta_1 + a^2\theta_3 + \ldots\right) - a\left(\theta_2 + a^2\theta_4 + \ldots\right) \leq x\right\}$$

$$= P\left\{\left(\theta_2 + a^2\theta_4 + \ldots\right) \geq \frac{\left(\theta_1 + a^2\theta_3 + \ldots\right) - x}{a}\right\}$$

$$= \int_0^x dF_s(y)\left[1 - F_s\left(\frac{y-x}{a}\right)\right],$$

where s is the sum of the series

$$s = \sum_{k=1}^{\infty} a^{k+2}\theta_k.$$

Suppose θ_1 has the uniform distribution on $[0,1]$, then the cdf $F_s(x) = F_{\nu}(x)$, where $F_{\nu}(x)$ is calculated in lemma 5.4.

In the case where θ_1 has the Erlang-2 distribution, the cdf $F_s(x) = F_{\eta}(x)$, where $F_{\eta}(x)$ is calculated in lemma 5.5.

5.3. Differential and integral equations for jump random motions

In this section, we obtain a differential equation for the characteristic function of random jump motion on the line, where the direction alternations and random jumps occur according to the renewal epochs of the Erlang distribution. We also study random jump motion in higher dimensions and we obtain a renewal-type equation for the characteristic function of the process. In the three-dimensional case, we obtain the telegraph-type differential equation for jump random motion, where the direction alternations and random jumps occur according to the renewal epochs of the Erlang-2 distribution.

In 1951 and 1956, Goldstein and Kac solved the problem of a one-dimensional random motion driven by a Poisson process, i.e. defined by the movement of a particle at a constant velocity v traveling in some direction a random distance drawn from an exponential distribution. After that, the direction of the particle changed in the opposite direction under the same stochastic conditions. The problem can be seen as a random motion governed by a switching Poisson process with alternating directions and having exponentially distributed sojourn times. Goldstein and Kac found that the solution of this problem satisfies the one-dimensional telegraph-type equation, which is similar to the Heaviside telegraph equation for wave propagation in transmission lines. Namely,

$$\frac{\partial^2 f(t,x)}{\partial t^2} + 2\lambda\frac{\partial}{\partial t}f(t,x) = v^2\frac{\partial^2 f(t,x)}{\partial x^2}, \qquad [5.45]$$

and it is called the Goldstein–Kac telegraph equation. Many variations of this basic idea have been approached by many researchers worldwide, and in this section we focus on the important case of the jump telegraph process on a line and in higher dimensions. The one-dimensional jump telegraph process, which is a generalization of the telegraph process, was introduced and studied (including its applications in financial market theory) by Ratanov (2007, 2010), and Ratanov and Melnikov (2008). Some limit theorems for the jump telegraph process on a line were studied by Di Crescenzo and Martinucci (2013). The jump telegraph process with random jumps was introduced and studied by Di Crescenzo *et al.* (2013). Some previous results regarding Erlang distributions in the telegraph process context can be found in Di Crescenzo (2001), Pogorui and Rodríguez-Dagnino (2005a) and Pogorui and Rodríguez-Dagnino (2012), random motions in multidimensional spaces in De Gregorio (2012, 2014), connections between random motion and PDEs in physics in Garra and Orsingher (2014) and recent results on jump-telegraph processes governed by alternating fractional Poisson process in Di Crescenzo and Meoli (2018).

In this section, we obtain differential equations for the one-dimensional jump telegraph process in the case of Erlang distributed switching times for velocities, and arbitrary distributed random jumps. We also study the jump telegraph process in $\mathbb{R}^n$, $n > 1$, and we obtain the Volterra integral equation of convolution type for the characteristic function of the process. As a particular case, we consider a differential equation for the jump process in $\mathbb{R}^3$ and study its solution.

5.3.1. *The Erlang jump telegraph process on a line*

Let us consider the random motion of a particle on the line described in the following manner: the particle moves according to one of two velocities $c > 0$ or $-v < 0$. Starting from the point $x_0 \in \mathbb{R}$, the particle moves with velocity c during the random time $\tau_1 = \xi_{\lambda_1} + \ldots + \xi_{\lambda_n}$, $n \geq 1$, where ξ_{λ_i} is exponentially distributed with rate $\lambda_i > 0$ and random variables $\{\xi_{\lambda_i}, i = 1, \ldots, n\}$. Then, the particle jumps a random length η_1 and continues its motion with velocity $-v$ during the random time $\tau_2 = \xi_{\mu_1} + \ldots + \xi_{\mu_m}$, $m \geq 1$, where ξ_{μ_i} is exponentially distributed with rate $\mu_i > 0$. Furthermore, the particle jumps a random length η_2 and moves with velocity c, and so on. It is assumed that random variables $\{\eta_1, \eta_2, \xi_{\lambda_i}, \xi_{\mu_j}, i = 1, \ldots, n, j = 1, \ldots, m\}$ are mutually independent. Thus, after even renewal events the particle has velocity c and after odd events velocity $-v$.

The motion of this particle can be described by using the theory of random evolutions in the following form. Denote by $\psi(t)$, $t \geq 0$, the alternating semi-Markov process on the phase (or state) space $T = \{1, 2\}$, with sojourn time τ_i at state $i \in T$, and with transition probabilities matrix of the embedded Markov chain

$$P = \begin{pmatrix} 0 & 1 \\ 1 & 0 \end{pmatrix}.$$

Denote by $y(t)$, $t \geq 0$, the position of the particle at time t.

Let us introduce the following function C_1 on T

$$C_1(y) = \begin{cases} c, & \text{if} \quad y = 1, \\ -v, & \text{if} \quad y = 2. \end{cases} \qquad [5.46]$$

Thus, the process $y(t)$ is a random evolution in the semi-Markov medium $\psi(t)$ satisfying the following equation (Korolyuk and Swishchuk 1995b):

$$y(t) = x_0 + \int_0^t C_1(\psi(s))\, ds + \sum_{k=1}^{\nu(t)} \eta_k, \qquad [5.47]$$

where $\nu(t)$ is the number of renewal events of $\psi(t)$, and η_k has the same distribution as η_1 for odd k, and η_k has the same distribution as η_2 for even k. In addition, the random variables $\eta_1, \eta_2, \ldots$ are mutually independent. This implies that $\left|\sum_{k=1}^{\nu(t)} \eta_k\right| < +\infty$ almost surely for any $t \geq 0$.

Now, let us show that the evolution $y(t)$ can be represented as an evolution driven by a Markov process. Denote by $\xi(t)$, $t \geq 0$, a Markov chain in the phase space $E = \{1, 2, \ldots, n, n+1, \ldots, n+m\}$ with the infinitesimal operator $Q = q[P - I]$, where $q = \operatorname{diag}(\lambda_1, \lambda_2, \ldots, \lambda_n, \mu_1, \ldots, \mu_m)$ and P is the following $(n+m) \times (n+m)$ transition probabilities matrix:

$$P = \|p_{ij}\| = \begin{pmatrix} 0 & 1 & 0 & \ldots & 0 \\ 0 & 0 & 1 & \ldots & 0 \\ \vdots & \vdots & \vdots & \ddots & \vdots \\ 0 & 0 & 0 & \ldots & 1 \\ 1 & 0 & 0 & \ldots & 0 \end{pmatrix}.$$

We assume that $\mathrm{P}[\xi(0) = 1] = 1$.

Let us introduce the following function C on E:

$$C(i) = \begin{cases} c, & \text{if} \quad i = 1, \ldots, n, \\ -v, & \text{if} \quad i = n+1, \ldots, n+m \end{cases}$$

and consider the equation

$$x(t) = x_0 + \int_0^t C(\xi(s))\,ds + \sum_{k=1}^{n(t)} \alpha_k, \qquad [5.48]$$

where $n(t)$ is the number of renewal events of $\xi(t)$, and $\alpha_{kn+(k-1)m}$, $k = 1, 2, \ldots$ has the same distribution as η_1, and α_{km+kn}, $k = 1, 2, \ldots$ has the same distribution as η_2, and $\alpha_k = 0$ in other cases. We suppose that there exists the pdf $g_{\eta_i}(y)$, $i = 1, 2$, and we assume $g_{\alpha_i}(y) = \delta(y)$ for $\alpha_i = 0$, where $\delta(\cdot)$ is the delta function.

Equation [5.48] describes a random evolution in Markov media $\xi(t)$, $t \geq 0$. In addition, it is easily verified that $\mathrm{P}[y(t) = x(t)] = 1$ for each $t \geq 0$. Thus, we reduce the semi-Markov case described by equation [5.47] to the Markov case described by equation [5.48].

In the rest of this section, we will study the process $x(t)$, which is described by equation [5.48], and we will consider the two-component process $\zeta(t) = (x(t), \xi(t))$ in the phase space $Z = \mathbb{R} \times E$, where $\mathbb{R} = (-\infty, +\infty)$, $E = \{1, \ldots, n, n+1, \ldots, n+m\}$.

It is well-known (Korolyuk and Korolyuk 1999; Korolyuk and Swishchuk 1995b) that the process $\zeta(t) = (x(t), \xi(t))$ has the following infinitesimal operator:

$$A\varphi(x,i) = C(x,i)\frac{\partial}{\partial x}\varphi(x,i) + \lambda_i(P\varphi(x,i) - \varphi(x,i)) + A_0\varphi(x,i),$$

where $\varphi(x,i)$ is a differential function with respect to $x \in \mathbb{R}$ for all $i \in E$ and such that there exists $\int_{-\infty}^{\infty} g_{\alpha_i}(y)\varphi(x+y,j)\,dy$ for any $x \in \mathbb{R}$, $i \in E$, $P\varphi(x,i) = \sum_{j\in E} p_{ij}\varphi(x,j)$ and the operator A_0 is of the following form, for $i = n$ or $i = n+m$,

$$A_0\varphi(x,i) = \lambda_i\left(\sum_{j\in E} p_{ij}\int_{-\infty}^{\infty} g_{\alpha_i}(y)\left[\varphi(x+y,j) - \varphi(x,i)\right]dy\right)$$

and for other $i \in E$

$$A_0\varphi(x,i) = 0.$$

Let us write the operator A in more detail:

$$A\varphi(x,1) = c\frac{\partial}{\partial x}\varphi(x,1) + \lambda_1(\varphi(x,2) - \varphi(x,1)),$$

$$A\varphi(x,2) = c\frac{\partial}{\partial x}\varphi(x,2) + \lambda_2(\varphi(x,3) - \varphi(x,2)),$$

$$\vdots$$

$$A\varphi(x,n) = c\frac{\partial}{\partial x}\varphi(x,n) + \lambda_n(\varphi(x,n+1) - 2\varphi(x,n))$$

$$+\lambda_n \int_{-\infty}^{\infty} g_{\eta_1}(y)\,\varphi(x+y,n+1)\,dy,$$

$$A\varphi(x,n+1) = -v\frac{\partial}{\partial x}\varphi(x,n+1) + \mu_1(\varphi(x,n+2) - \varphi(x,n+1)),$$

$$\vdots$$

$$A\varphi(x,n+m) = -v\frac{\partial}{\partial x}\varphi(x,n+m) + \mu_m(\varphi(x,1) - 2\varphi(x,n+m))$$

$$+\mu_m \int_{-\infty}^{\infty} g_{\eta_2}(y)\,\varphi(x+y,1)\,dy.$$

Consider $f(t,x,k)\,dx = P\{x \le x(t) \le x+dx, \xi(t) = k\}$, $k \in E$. The function $f(t,x,k)$ satisfies the Kolmogorov backward equation

$$\frac{\partial f(t,x,k)}{\partial t} = A f(t,x,k), \qquad [5.49]$$

with the initial condition

$$\sum_{k\in E} f(0,x,k) = \delta(x).$$

Passing, in equation [5.49], into characteristic functions

$$H_k(t,\omega) = \int_{-\infty}^{\infty} e^{i\omega x} f(t,x,k)dx,$$

we obtain

$$\frac{\partial H_1(t,\omega)}{\partial t} = -i\omega c H_1(t,\omega) + \lambda_1(H_2(t,\omega) - H_1(t,\omega)),$$

$$\frac{\partial H_2(t,\omega)}{\partial t} = -i\omega c H_2(t,\omega) + \lambda_2(H_3(t,\omega) - H_2(t,\omega)),$$

$$\vdots$$

$$\frac{\partial H_n(t,\omega)}{\partial t} = -i\omega c H_n(t,\omega) + \lambda_n(H_{n+1}(t,\omega) - 2H_n(t,\omega))$$

$$+\lambda_n \hat{g}_{\eta_1}(-\omega)\,H_{n+1}(t,\omega),$$

$$\frac{\partial H_{n+1}(t,\omega)}{\partial t} = i\omega v H_{n+1}(t,\omega) + \mu_1 (H_{n+2}(t,\omega) - H_{n+1}(t,\omega)),$$

$$\vdots$$

$$\frac{\partial H_{n+m}(t,\omega)}{\partial t} = i\omega v H_{n+m}(t,\omega) + \mu_m (H_1(t,\omega) - 2H_{n+m}(t,\omega))$$

$$+\mu_m \hat{g}_{\eta_2}(-\omega) H_1(t,\omega). \qquad [5.50]$$

Let us write equation [5.50] in the matrix form

$$M\mathbf{H}(t,\omega) = 0,$$

where $\mathbf{H}(t,\omega) = (H_1(t,\omega), H_2(t,\omega), \ldots, H_{n+m}(t,\omega))^\top$ and the matrix M is as follows:

$$M = \begin{pmatrix} \frac{\partial}{\partial t} + i\omega c + \lambda_1 & -\lambda_1 & 0 & \cdots & 0 \\ 0 & \frac{\partial}{\partial t} + i\omega c + \lambda_2 & -\lambda_2 & \cdots & 0 \\ \vdots & \vdots & \vdots & \ddots & \vdots \\ -\mu_m - \mu_m \hat{g}_{\eta_2}(-\omega) & \cdots & 0 & \cdots & \frac{\partial}{\partial t} - i\omega v + 2\mu_m \end{pmatrix}$$

Consider the function $H(t,\omega) = \sum_{i=1}^{n+m} H_i(t,\omega)$. It is easily seen that $H(t,\omega)$ is the characteristic function of $f(t,x)$, where $f(t,x) = \sum_{i=1}^{n+m} f(t,x,i)$. The function $f(t,x)$ is the pdf of the position $x(t)$ of the particle at time t.

THEOREM 5.4.– The function $H(t,\omega)$ satisfies the following differential equation:

$$\prod_{k=1}^{n-1} \left(\frac{\partial}{\partial t} + i\omega c + \lambda_k \right) \prod_{l=1}^{m-1} \left(\frac{\partial}{\partial t} - i\omega v + \mu_l \right)$$

$$\times \left(\frac{\partial}{\partial t} + i\omega c + 2\lambda_n \right) \left(\frac{\partial}{\partial t} - i\omega v + 2\mu_m \right) H(t,\omega)$$

$$-(1+\hat{g}_{\eta_1}(-\omega))(1+\hat{g}_{\eta_2}(-\omega)) \prod_{i=1}^{n} \lambda_i \prod_{j=1}^{m} \mu_j H(t,\omega) = 0. \qquad [5.51]$$

DEMONSTRATION.– It is well-known (Pogorui and Rodríguez-Dagnino 2005a) that $H(t,\omega)$ satisfies the equation

$$(\det(M)) H(t,\omega) = 0, \qquad [5.52]$$

where $\det(M)$ is the determinant of the matrix M. It is easily verified that

$$\det(M) = \prod_{k=1}^{n} \left(\frac{\partial}{\partial t} + i\omega c + \lambda_k \right) \prod_{l=1}^{m} \left(\frac{\partial}{\partial t} - i\omega v + \mu_l \right)$$

$$\times\left(\frac{\partial}{\partial t}+i\omega c+2\lambda_n\right)\left(\frac{\partial}{\partial t}-i\omega v+2\mu_m\right)$$

$$-\left(1+\hat{g}_{\eta_1}(-\omega)\right)\left(1+\hat{g}_{\eta_2}(-\omega)\right)\prod_{i=1}^{n}\lambda_i\prod_{j=1}^{m}\mu_j.$$

□

Consider the case where $c=v$, $n=m$, $\lambda_i=\mu_i=\lambda$. In this case, equation [5.51] has the following form:

$$\left(\frac{\partial^2}{\partial t^2}+2\lambda\frac{\partial}{\partial t}+\omega^2c^2+\lambda^2\right)^{n-1}\left(\frac{\partial^2}{\partial t^2}+4\lambda\frac{\partial}{\partial t}+\omega^2c^2+4\lambda^2\right)H(t,\omega)$$

$$-\left(1+\hat{g}_{\eta_1}(-\omega)\right)\left(1+\hat{g}_{\eta_2}(-\omega)\right)\lambda^{2n}H(t,\omega)=0. \qquad [5.53]$$

Without loss of generality, we assume that $\mathrm{P}(\xi(0)=1)=1$, i.e. the particle starts moving with velocity c. Therefore, $H(0,\omega)=H_1(0,\omega)=1$. It follows from the first and the last equations of the set of [5.50] that (all derivatives are taken with respect to t)

$$H'(0,\omega)=H_1'(0,\omega)+H_{n+m}'(0,\omega)=-i\omega c+\lambda\hat{g}_{\eta_2}(-\omega).$$

Similarly, we can calculate $H^{(n)}(0,\omega)$ for any $n\in\mathbb{N}$.

Now, the pdf $f(t,x)$ can be obtained through the inverse Fourier transform applied to equation [5.53]. Hence,

$$\left(\frac{\partial^2}{\partial t^2}+2\lambda\frac{\partial}{\partial t}-c^2\frac{\partial^2}{\partial x^2}+\lambda^2\right)^{n-1}\left(\frac{\partial^2}{\partial t^2}+4\lambda\frac{\partial}{\partial t}-c^2\frac{\partial^2}{\partial x^2}+4\lambda^2\right)f(t,x)$$

$$-\lambda^{2n}f(t,x)$$

$$+\lambda^{2n}\int_{-\infty}^{\infty}\left(g_{\eta_1}(-x+u)+g_{\eta_2}(-x+u)-g_{\eta_1+\eta_2}(-x+u)\right)f(t,u)\,du=0,$$

with initial conditions $f(0,x)=\delta(x)$, $f'(0,x)=-c\delta'(x)+\lambda g_{\eta_2}(-x)$ and so on, where $\left(\frac{\partial^2}{\partial t^2}+2\lambda\frac{\partial}{\partial t}-c^2\frac{\partial^2}{\partial x^2}+\lambda^2\right)^m$ is a formal m-times product of the differential operator $\frac{\partial^2}{\partial t^2}+2\lambda\frac{\partial}{\partial t}-c^2\frac{\partial^2}{\partial x^2}+\lambda^2$, and it is assumed that $\left(\frac{\partial^2}{\partial t^2}\right)^m=\frac{\partial^{2m}}{\partial t^{2m}}$.

5.3.2. *Examples*

5.3.2.1. *Exponential case*

In the case where $n = 1$, equation [5.53] can be written in the following form:

$$\left[\frac{\partial^2}{\partial t^2} + 4\lambda\frac{\partial}{\partial t} + \omega^2 c^2 + 4\lambda^2 - (1 + \hat{g}_{\eta_1}(-\omega))(1 + \hat{g}_{\eta_2}(-\omega))\right]$$

$$\times H(t, \omega) = 0, \qquad [5.54]$$

with the initial conditions $H(0, \omega) = 1$, $H'(0, \omega) = -i\omega c + \lambda\hat{g}_{\eta_2}(-\omega)$.

By solving equation [5.54], we obtain the characteristic function

$$H(t, \omega) = e^{-2\lambda t}\left(\cosh(A(\omega)t) + \frac{2\lambda - i\omega c + \lambda\hat{g}_{\eta_2}(-\omega)}{A(\omega)}\sinh(A(\omega)t)\right),$$

where $A(\omega) = \sqrt{\lambda^2(1 + \hat{g}_{\eta_1}(-\omega) + \hat{g}_{\eta_2}(-\omega) + \hat{g}_{\eta_1}(-\omega)\hat{g}_{\eta_2}(-\omega)) - \omega^2 c^2}$.

5.3.2.2. *Erlang-2 case*

In the case where $n = 2$, equation [5.53] can be written in the following form:

$$\left(\frac{\partial^2}{\partial t^2} + 2\lambda\frac{\partial}{\partial t} + \omega^2 c^2 + \lambda^2\right)\left(\frac{\partial^2}{\partial t^2} + 4\lambda\frac{\partial}{\partial t} + \omega^2 c^2 + 4\lambda^2\right)H(t, \omega)$$

$$-\lambda^4(1 + \hat{g}_{\eta_1}(-\omega))(1 + \hat{g}_{\eta_2}(-\omega))H(t, \omega) = 0,$$

with the initial conditions $H(0, \omega) = H_1(0, \omega) = 1$;

$$\frac{\partial H(0, \omega)}{\partial t} = \frac{\partial H_1(0, \omega)}{\partial t} + \frac{\partial H_4(0, \omega)}{\partial t} = -i\omega c + \lambda\hat{g}_{\eta_2}(-\omega);$$

$$\frac{\partial^2 H(0, \omega)}{\partial t^2} = \frac{\partial^2 H_1(0, \omega)}{\partial t^2} + \frac{\partial^2 H_3(0, \omega)}{\partial t^2} + \frac{\partial^2 H_4(0, \omega)}{\partial t^2}$$
$$= -\omega^2 c^2 - \lambda^2 + \lambda^2\hat{g}_{\eta_2}(-\omega) - 2i\omega c\lambda\hat{g}_{\eta_2}(-\omega);$$

$$\frac{\partial^3 H(0, \omega)}{\partial t^3} = \frac{\partial^3 H_1(0, \omega)}{\partial t^3} + \frac{\partial^3 H_2(0, \omega)}{\partial t^3} + \frac{\partial^3 H_3(0, \omega)}{\partial t^3} + \frac{\partial^3 H_4(0, \omega)}{\partial t^3}$$
$$= 6ic\lambda^2\omega\hat{g}_{\eta_2}(-\omega) - 3c^2\lambda\omega^2\hat{g}_{\eta_2}(-\omega) + 3ic\lambda^2\omega + \lambda^3(\hat{g}_{\eta_1}(-\omega)$$
$$+5\hat{g}_{\eta_2}(-\omega) + 3) + ic^3\omega^3.$$

Thus, the characteristic function is given by

$$H(t, \omega) = e^{-\frac{3\lambda}{2}t}(C_1\cosh(A(\omega)t) + C_2\cosh(B(\omega)t)$$
$$+C_3\sinh(A(\omega)t) + C_4\sinh(B(\omega)t)),$$

where

$$A(\omega) = \sqrt{L - 4\lambda\sqrt{\lambda^2 \left(1 + \hat{g}_{\eta_1}(-\omega) + \hat{g}_{\eta_2}(-\omega) + \hat{g}_{\eta_1}(-\omega)\,\hat{g}_{\eta_2}(-\omega)\right) + 2\omega^2 c^2}},$$

$$B(\omega) = \sqrt{L + 4\lambda\sqrt{\lambda^2 \left(1 + \hat{g}_{\eta_1}(-\omega) + \hat{g}_{\eta_2}(-\omega) + \hat{g}_{\eta_1}(-\omega)\,\hat{g}_{\eta_2}(-\omega)\right) + 2\omega^2 c^2}}$$

where $L = \lambda^2 - 4\omega^2 c^2$, and the coefficients C_i, $i = 1, 2, 3, 4$ can be calculated by using initial conditions.

5.4. Estimation of the number of level crossings by the telegraph process

It is well-known that for almost all sample paths of Brownian motion $W(t)$, $(W(0) = 0)$ (that is, for a set of sample paths having probability one), for all $\delta > 0$, the path has infinitely many zeros in the interval $(0, \delta)$. Under Kac's condition, the telegraph process weakly converges to the Wiener process. In this section, we estimate the number of level crossings by the telegraph process and, passing to the limit in Kac's condition, we obtain the estimation for a level crossing by the Wiener process.

Let $\{\xi(t), t \geq 0\}$ be a Markov process on the phase space $\{0, 1\}$ with the generative matrix

$$Q = \lambda \begin{pmatrix} -1 & 1 \\ 1 & -1 \end{pmatrix}.$$

DEFINITION 5.2.– *$x(t)$ is the telegraph process if*

$$\frac{d}{dt}x(t) = v(-1)^{\xi(t)}, \ v = const > 0, \qquad x(0) = x_0. \qquad [5.55]$$

In the sequel, we need some auxiliary results. The explicit form for the distribution of the first passage time of a level by a telegraph processes on the line.

For a fixed l, let us define that $\Delta(t) = l - x(t)$. We assume that $z = l - x_0 > 0$.

Suppose $\xi(0) = k$ and define

$$\tau_k(z) = \inf\{t \geq 0 : \Delta(t) = 0\}, \ k \in \{0, 1\}.$$

It is easily seen that $\tau_k(z)$ is the first passage time of the level l by $x(t)$ assuming $\xi(0) = k$.

Denote by $f_k(t, z)\,dt = P(\tau_k(z) \in dt)$ the probability density function of $\tau_k(z)$.

THEOREM 5.5.– For $t \geq \frac{z}{v}$

$$f_0(t,z) = e^{-\lambda t}\delta(z - vt) + \frac{z\lambda}{v^2} e^{-\lambda t} \frac{I_1\left(\frac{\lambda}{v}\sqrt{v^2t^2 - z^2}\right)}{\sqrt{v^2t^2 - z^2}}, \quad [5.56]$$

$$f_1(t,z) = \frac{e^{-\lambda t}}{t - \frac{z}{v}} I_1\left(\lambda\left(t - \frac{z}{v}\right)\right) \quad [5.57]$$

$$+ z\lambda \int_{z/v}^{t} \frac{e^{-\lambda t} I_1(\lambda(t-u)) I_1\left(\frac{\lambda}{v}\sqrt{v^2u^2 - z^2}\right)}{(t-u)\sqrt{v^2u^2 - z^2}} du,$$

where $I_1(\cdot)$ is the modified Bessel function of the first kind.

PROOF.– Let us consider the Laplace transforms of $\tau_k(z)$, $k \in \{0,1\}$.

$$\varphi_k(s,z) = \mathbf{E}\left[e^{-s\tau_k(z)}\right], \quad s > 0.$$

By using the renewal theory, we can obtain the following system of integral equations for these Laplace transforms:

$$\varphi_0(s,z) = e^{-\frac{s+\lambda}{v}z} + \frac{\lambda}{v}\int_0^z e^{-\frac{s+\lambda}{v}u}\varphi_1(s, z-u)\, du$$

$$= e^{-\frac{s+\lambda}{v}z} + \frac{\lambda}{v} e^{-\frac{s+\lambda}{v}z}\int_0^z e^{\frac{s+\lambda}{v}u}\varphi_1(s,u)\, du,$$

$$\varphi_1(s,z) = \frac{\lambda}{v} e^{\frac{s+\lambda}{v}z}\int_z^\infty e^{-\frac{s+\lambda}{v}u}\varphi_0(s,u)\, du.$$

These equations are then differentiated to obtain the following system:

$$\frac{\partial}{\partial z}\varphi_0(s,z) = -\frac{s+\lambda}{v}\varphi_0(s,z) + \frac{\lambda}{v}\varphi_1(s,u)$$

$$\frac{\partial}{\partial z}\varphi_1(s,z) = \frac{s+\lambda}{v}\varphi_1(s,z) - \frac{\lambda}{v}\varphi_0(s,u).$$

It is well-known (Pogorui and Rodríguez-Dagnino 2005a) that $\varphi_0(s,z)$ and $\varphi_1(s,z)$ satisfy the following equation:

$$\det\begin{pmatrix} \frac{\partial}{\partial z} + \frac{s+\lambda}{v} & -\frac{\lambda}{v} \\ \frac{\lambda}{v} & \frac{\partial}{\partial z} - \frac{s+\lambda}{v} \end{pmatrix} f(z) = 0.$$

By calculating the determinant, we have

$$\frac{\partial^2}{\partial z^2} f(z) - \frac{s^2 + 2\lambda s}{v^2} f(z) = 0.$$

and the solution of this equation can be written as

$$f(z) = C_1 e^{\sqrt{s^2+2\lambda s}\frac{z}{v}} + C_2 e^{-\sqrt{s^2+2\lambda s}\frac{z}{v}}.$$

The constants obtained from the system of integral equations yields

$$\varphi_0(s,z) = e^{-\frac{z}{v}\sqrt{s^2+2\lambda s}}, \tag{5.58}$$

and

$$\varphi_1(s,z) = \frac{s+\lambda-\sqrt{s^2+2\lambda s}}{\lambda} e^{-\frac{z}{v}\sqrt{s^2+2\lambda s}}. \tag{5.59}$$

The inverse Laplace transform of $\varphi_{(0,1)}(s,z)$ with respect to s yields the following pdf (Korn and Korn 2000, p. 239, formula 88):

$$\begin{aligned} f_0(t,z) &= \mathcal{L}^{-1}\left(e^{-\frac{z}{v}\sqrt{s^2+2\lambda s}}, s\right) \\ &= e^{-\lambda t}\delta(z-vt) + z\lambda e^{-\lambda t}\frac{I_1\left(\frac{\lambda}{v}\sqrt{v^2t^2-z^2}\right)}{\sqrt{v^2t^2-z^2}}, \end{aligned} \tag{5.60}$$

$$t \geq \frac{z}{v}.$$

Hence,

$$P(\tau_0(z) \in dt) = e^{-\lambda t}\delta(z-vt)\,dt + z\lambda e^{-\lambda t}\frac{I_1\left(\frac{\lambda}{v}\sqrt{v^2t^2-z^2}\right)}{\sqrt{v^2t^2-z^2}}dt, \tag{5.61}$$

where $t \geq \frac{z}{v}$.

It follows from (Korn and Korn 2000, p. 237), we have

$$\mathcal{L}^{-1}\left(\frac{s+\lambda-\sqrt{s^2+2\lambda s}}{\lambda}, s\right) = \frac{e^{-\lambda t}}{t} I_1(\lambda t).$$

Therefore,

$$f_1(t,0) = \frac{e^{-\lambda t}}{t} I_1(\lambda t).$$

Taking into account equations [5.60] and [5.61], we have for $t \geq \frac{z}{v}$

$$\begin{aligned} f_1(t,z) &= \int_{z/v}^{t} \frac{e^{-\lambda(t-u)}}{t-u} I_1(\lambda(t-u)) \\ &\times \left[e^{-\lambda u}\delta(z-vu) + z\lambda e^{-\lambda u}\frac{I_1\left(\frac{\lambda}{v}\sqrt{v^2u^2-z^2}\right)}{\sqrt{v^2u^2-z^2}}\right] du \\ &= \frac{e^{-\lambda t}}{t-\frac{z}{v}} I_1\left(\lambda\left(t-\frac{z}{v}\right)\right) + z\lambda\int_{z/v}^{t} \frac{e^{-\lambda t} I_1(\lambda(t-u))\, I_1\left(\frac{\lambda}{v}\sqrt{v^2u^2-z^2}\right)}{(t-u)\sqrt{v^2u^2-z^2}} du. \end{aligned}$$

5.4.1. *Estimation of the number of level crossings for the telegraph process in Kac's condition*

Denote by $C_k(t,z)$a number of intersections of level z by the particle $x(t)$ during $(0,t)$, $t>0$ assuming $\xi(0)=k\in\{0,1\}$.

Consider the following renewal function $H_k(t,z)=\mathbf{E}\left[C_k(t,z)\right]$.

Let us consider the following so-called Kac's condition (or the hydrodynamic limit). Denote, by $\lambda=\varepsilon^{-2}$, $v=c\varepsilon^{-1}$; as $\varepsilon>0$; i.e. $\lambda>0$ and $v>0$; such that $\frac{v^2}{\lambda}>c^2$.

It was proved in Kac (1974) that, under Kac's condition, the telegraph process $x(t)$ weakly converges to the Wiener process $W(t)\sim N(0,ct)$

THEOREM 5.6.– Under Kac's condition, we have

$$\lim_{\lambda\to\infty}\frac{H_1(t,0)}{\sqrt{\lambda}}=\lim_{v\to\infty}\frac{cH_1(t,0)}{v}=\lim_{\varepsilon>0}\varepsilon H_1(t,0)=\sqrt{2t/\pi}.$$

PROOF.– By using the Laplace transform for the general renewal function (Cox 1970, p. 56, formula 2), it follows from equation [5.59] that the Laplace transform $\hat{H}_1(s,0)=\mathcal{L}(H_1(t,0),t)$ of $H_1(t,0)$ with respect to t has the form

$$\hat{H}_1(s,0)=\frac{1}{s}\sum_{k=0}^{\infty}\left(\frac{s+\lambda-\sqrt{s^2+2\lambda s}}{\lambda}\right)^k=\frac{\lambda}{s\sqrt{s^2+2\lambda s}-s^2}.$$

It is easily verified that

$$\begin{aligned}\mathbf{E}\left[C_1(t,0)\right]&=\mathcal{L}^{-1}\left(\frac{\lambda}{s\sqrt{s^2+2\lambda s}-s^2}\right)\\&=\frac{1}{2}+\left(\left(\frac{1}{2}+\lambda t\right)I_0(2\lambda t)+\lambda tI_1(2\lambda t)\right)e^{-\lambda t}.\end{aligned}$$

Hence,

$$\lim_{\lambda\to\infty}\frac{\mathbf{E}\left[C_1(t,0)\right]}{\sqrt{\lambda}}=\sqrt{2t/\pi}.$$

Taking into account $\lambda=\varepsilon^{-2}$, $v=c\varepsilon^{-1}$, we have

$$H_1(t,0)\sim t\sqrt{2/\pi}\sqrt{\lambda}=c\sqrt{2t/\pi v},\quad \text{as } v\to\infty.$$

Therefore, for a fixed $t>0$, the number of level crossings by the telegraph process under Kac's condition goes to infinity as the velocity v does.

Denote as

$$F_1(x) = \int_0^x f_1(t,0)dt = \int_0^x \frac{e^{-\lambda t}}{t} I_1(\lambda t)\, dt.$$

THEOREM 5.7.–

$$P\left\{\left(1 - \int_0^{\lambda x} \frac{e^{-u}}{u} I_1(u)\, du\right) C_1(x,0) \geq \frac{3}{\sqrt{y}}\right\} \to G_{1/2}(y) \text{ as } \lambda \to \infty,$$

where $G_{1/2}(y)$ is the one-sided stable distribution satisfying the condition $y^{1/2}\left[1 - G_{1/2}(y)\right] \to 3$ as $y \to \infty$ (Feller 1971).

PROOF.– It is easily seen that

$$\begin{aligned}\lim_{x\to\infty} (1 - F_1(x))\sqrt{x} &= \lim_{x\to\infty}\left(1 - \int_0^{\lambda x} \frac{e^{-2\lambda t}}{t} I_1(2\lambda t)\, dt\right)\sqrt{x} \\ &= 2\lim_{x\to\infty} e^{-2\lambda x} x^{1/2} I_1(2\lambda x) = \frac{2}{\sqrt{\pi}}.\end{aligned}$$

Therefore, $L(x) = (1 - F_1(x))\sqrt{x}$ is slowly varying and

$$1 - F_1(x) = x^{-1/2} L(x).$$

Since $F_1(x) = \int_0^x \frac{e^{-\lambda t}}{t} I_1(\lambda t)\, dt = \int_0^{\lambda x} \frac{e^{-u}}{u} I_1(u)\, du$ and $\lambda \to \infty$ we have $F_1(x) = F(s) = s^{-1/2} L(s)$, where for a fixed $x > 0$, $s = \lambda x \to \infty$. By using the result in (Feller 1971, Chapter XI.5, p. 373), we obtain

$$P\left\{\left(1 - \int_0^{\lambda x} \frac{e^{-u}}{u} I_1(u)\, du\right) C_1(x,0) \geq \frac{3}{\sqrt{y}}\right\} \to G_{1/2}(y) \text{ as } \lambda \to \infty,$$

where $G_{1/2}(y)$ is the one-sided stable distribution satisfying the condition $y^{1/2}\left[1 - G_{1/2}(y)\right] \to 3$ as $y \to \infty$.

Therefore, under Kac's condition, the number of level crossings by the telegraph process $C_1(x,0)$ is of the order of magnitude $\sqrt{\lambda} = v$.

References

Albeverio, S., Korolyuk, V., Samoilenko, I. (2009). Asymptotic expansion of semi-Markov random evolutions. *Stochastics*, 81(5), 477–502.

Anick, D., Mitra, D., Sondhi, M.M. (1982). Stochastic theory of a data-handling system. *Bell System Technical Journal*, 61(8), 1871–1894.

Anisimov, V.V. (1977). Switched processes. *Kibernetika*, 4, 111–115.

Arratia, R.A. (1979). Coalescing Brownian motions on the line. PhD Thesis, University of Wisconsin, Madison.

Balakrishnan, V., Van den Broeck, C., Hänggi, P. (1988). First-passage of non-Markovian processes: The case of a reflecting boundary. *Physical Review A*, 38, 4213–4222.

Barlow, R.E., Proschan, F., Hunter, L. (1965). *Mathematical Theory of Reliability*. John Wiley & Sons, New York.

Bateman, H. (1954). *Tables of Integral Transforms*, McGraw Hill, New York.

Beghin, L. and Orsingher, E. (2010a). Moving randomly amid scattered obstacles. *Stochastics*, 82, 201–229.

Beghin, L. and Orsingher, E. (2010b). Poisson-type processes governed by fractional and higher-order recursive differential equations. *Electronic Journal of Probability*, 15, 684–709.

Bertoin, J. (2002). Self-attracting Poisson clouds in an expanding universe. *Communications in Mathematical Physics*, 232, 59–81.

Borovkov, A.A. and Mogulsky, A.A. (1980). On large deviation probabilities in topological spaces. II. *Siberian Mathematical Journal*, 21(5), 12–26.

Buzacott, J.A. (1967). Prediction of the efficiency of production systems without internal storage. *International Journal of Production Research*, 6(3), 173–188.

Cahoy, D.O. (2007). Fractional Poisson process in terms of α-stable densities. PhD Thesis, Case Western University, Cleveland.

Cherkesov, G.N. (1974). *Reliability of Engineering Systems with Temporal Redundancy*. Sovetskoe Radio, Moscow.

Corlat, A.N., Kuznetsov, V.N., Novikov, M.M., Turbin, A.F. (1991). *Semi-Markov Models of Renewal and Queuing Systems*. Shtiintsa, Chisinau.

Cox, D.R. (1970). *Renewal Theory*. Methuen & Co., London.

Cox J. and Ross R. (1976). Valuation of options for alternative stochastic processes. *Journal of Financial Economics*, 3(1–2), 145–166.

Dawson, D. (1993). *Measure-valued Markov Processes*. Springer-Verlag, Berlin.

Dawson, D.A. and Gärtner, J. (1994). Multilevel large deviations and interacting diffusions. *Probability Theory and Related Fields*, 98, 423–487.

Da Fonseca J., Ielpo F., Grasselli M. (2009). Hedging (co)variance risk with variance swaps. Available at: http://ssrn.com/abstract=1341811.

De Gregorio, A. (2012). On random flights with non-uniformly distributed directions. *Journal of Statistical Physics*, 147(2), 382–411.

De Gregorio, A. (2014). A family of random walks with generalized Dirichlet steps. *Journal of Mathematical Physics*, 55, 023302.

De Gregorio, A. and Orsingher, E. (2012). Flying randomly in $\mathbb{R}^d$ with Dirichlet displacements. *Stochastic Processes and their Applications*, 122, 676–713.

De Gregorio, A., Orsingher, E., Sakhno, L. (2005). Motions with finite velocity analyzed with order statistics and differential equations. *Theory of Probability and Mathematical Statistics*, 71, 63–79.

Di Crescenzo, A. (2001). On random motions with velocities alternating at Erlang-distributed random times. *Adv. Appl. Prob*, 33, 690–701.

Di Crescenzo, A. and Martinucci, B. (2010). A damped telegraph random process with logistic stationary distribution. *Journal of Applied Probability*, 47(1), 84–96.

Di Crescenzo, A. and Martinucci, B. (2013). On the generalized telegraph process with deterministic jumps. *Methodology and Computing in Applied Probability*, 15(1), 215–235.

Di Crescenzo, A. and Meoli, A. (2018). On a jump-telegraph process driven by an alternating fractional Poisson process. *Journal of Applied Probability*, 55(1), 94–111.

Di Crescenzo, A., Iuliano, A., Martinucci, B., Zacks, S. (2013). Generalized telegraph process with random jumps. *Journal of Applied Probability*, 50(2), 450–463.

Dineen, S. (2005). *Probability Theory in Finance: A Mathematical Guide to the Black-Scholes Formula*. American Mathematical Society, Rhode Island.

Donsker, M.D. and Varadhan, S.R.S. (1975). Large deviations for Markov processes and the asymptotic evaluation of certain Markov process expectations for large times. In *Probabilistic Methods in Differential Equations, Lecture Notes in Mathematics*, Volume 451, Pinsky, M.A. (eds). Springer, Berlin, Heidelberg. Available at: ttps://doi.org/10.1007/BFb0068580.

Dorogovtsev, A.A. (2007a). Fourier-Wiener transforms of functionals on Arratia flow. *Ukrainian Mathematical Bulletin*, 4(3), 333–354.

Dorogovtsev, A.A. (2007b). Measure-valued stochastic processes and flows. Institute Mathematics NAS of Ukraine, Kyiv.

Dorogovtsev, A.A. (2010). One Brownian stochastic flow. *Theory of Stochastic Processes*, 10, 21–25.

D'Ovidio, M., Orsingher, E., Toaldo, B. (2014). Fractional telegraph-type equations and hyperbolic Brownian motion. *Statistics & Probability Letters*, 89, 131–137.

Dynkin, E.B. (1991). *Markov Processes*. Springer-Verlag, Berlin, Heidelberg.

Elliott, R. and Swishchuk, A.V. (2007). Pricing options and variance swaps in Markov-modulated Brownian markets. In *Hidden Markov Models in Finance*, Mamon, R., Elliott, R. (eds), Springer, New York.

Elwalid, A.I. and Mitra, D. (1994). Statistical multiplexing with loss priorities in rate-based congestion control of high-speed networks. *IEEE Transactions on Communications*, 42(11), 2989–3002.

Feller, W. (1968). *An Introduction to Probability Theory and Its Applications*, Volume 1, 3rd edition John Wiley & Sons, New York.

Feller, W. (1971). *An Introduction to Probability Theory and Its Applications*, Volume 2, 2nd edition John Wiley & Sons, New York.

Foss, S. (2007). On exact asymptotics for a stationary sojourn time distribution in a tandem of queues for a class of light-tailed distributions. *Problems of Information Transmission*, 43(4), 93–108.

Franceschetti, M. (2007). When a random walk of fixed length can lead uniformly anywhere inside a hypersphere. *Journal of Theoretical Probability*, 20, 813–823.

Garra, R. and Orsingher, E. (2014). Random flights governed by Klein-Gordon-type partial differential equations. *Stochastic Processes and their Applications*, 124, 2171–2187.

Gikhman, I.I. and Skorokhod, A.V. (1973). *The Theory of Stochastic Processes II*. Nauka, Moscow.

Giraud, C. (2001). Clustering in a self-gravitating one-dimensional gas at zero temperature. *Journal of Statistical Physics*, 105, 585–604.

Giraud, C. (2005). Gravitational clustering and additive coalescence. *Stochastic Processes and their Applications*, 115, 1302–1322.

Girko, V.L. (1982). Limit theorems for products of random matrices with positive elements. *Teor. Veroyatnost. i Primenen.*, 27, 777–783.

Gnedenko, B. and Ushakov, I. (1995). *Probabilistic Reliability Engineering*. John Wiley & Sons, New York.

Goldstein, S. (1951). On diffusion by discontinuous movements and on the telegraph equation. *The Quarterly Journal of Mechanics and Applied Mathematics*, 4, 129–156.

Gorostiza, L.G. (1973). The central limit theorem for random motions of d-dimensional Euclidean space. *The Annals of Probability*, 1(4), 603–612.

Gorostiza, L.G. and Griego, R.J. (1979). Strong approximations of diffusion processes by transport processes. *Kyoto Journal of Mathematics*, 19(1), 91–103.

Gradshteyn, I.S. and Ryzhik, I.M. (1980). *Tables of Integrals, Sums, Series and Products*. Academic Press, San Diego.

Griego, R. and Hersh, R. (1969). Random evolutions, Markov chains, and systems of partial differential equations. *Proceedings of the National Academy of Sciences*, 62, 305–308.

Griego, R. and Hersh, R. (1971). Theory of random evolutions and applications to partial differential equations. *Transactions of the American Mathematical Society*, 156, 405–418.

Griego, R. and Korzeniowski, A. (1989). On principal eigenvalues for random evolutions. *Stochastic Analysis and Applications*, 7, 35–45.

Hersh, R. (1974). Random evolutions: A survey of results and problems. *Rocky Mountain Journal of Mathematics*, 4, 443–477.

Hersh, R. (2003). The birth of random evolutions. *The Mathematical Intelligencer*, 25(1), 53–60.

Hersh, R. and Papanicolaou, G. (1972). Non-commuting random evolutions, and an operator-valued Feynman-Kac formula. *Communications on Pure and Applied Mathematics*, 30, 337–367.

Hersh, R. and Pinsky, M. (1972). Random evolutions are asymptotically Gaussian. *Communications on Pure and Applied Mathematics*, 25, 33–44.

Iksanov, A.M. (2006). On the rate of convergence of a regular martingale related to the branching random walk. *Ukrainian Mathematical Journal*, 58(3), 326–342.

Iksanov, A.M. and Résler, U. (2006). Some moment results about the limit of a martingale related to the supercritical branching random walk and perpetuities. *Ukrainian Mathematical Journal*, 58(4), 451–471.

Jacod, J. and Shiryaev, A.N. (2010). *Limit Theorems for Stochastic Processes*. Springer-Verlag, Berlin.

Kabanov, Y. and Pergamenshchiskov, S. (2003). *Two-Scale Stochastic Systems: Asymptotic Analysis and Control*. Springer-Verlag, Berlin.

Kac, M. (1951). On some connections between probability theory and differential and integral equations. *Proceedings of the Fifth Berkeley Symposium on Mathematical Statistics and Probability*, 12, 189–215.

Kac, M. (1974). A stochastic model related to the telegrapher's equation. *Rocky Mountain Journal of Mathematics*, 4, 497–509.

Kartashov, N.V. (1985). Criteria for uniform ergodicity and strong stability of Markov chains with a common phase space. *Theory of Probability and Mathematical Statistics*, 30, 71–89.

Kato, T. (1980). *Perturbation Theory for Linear Operators*, 2nd edition Springer-Verlag, Berlin.

Kemeny, J.G. and Snell, J.L. (1960). *Finite Markov Chains*. Van Nostrand, New York.

Kertz, R. (1978). Random evolutions with underlying semi-Markov processes. *Publications of the Research Institute for Mathematical Sciences*, 14(3), 589–614.

Konarovskii, V.V. (2011). On infinite system of diffusing particles with coalescing. *Theory of Probability and Its Applications*, 55(1), 134–144.

Korn, G.A. and Korn, T.M. (2000). *Mathematical Handbook for Scientists and Engineers*, 2nd edition Dover Publications, New York.

Korolyuk, D.V. and Silvestrov, D.S. (1983). Time of reaching asymptotically receding domains for ergodic Markov chains [in Russian, with an English summary]. *Теория вероятностей и ее применения*, 28(2), 410–420.

Kornfeld, I.P., Ya, G.S., Fomin, S.V. (1982). *Ergodic Theory*. Springer-Verlag, New York.

Korolyuk, V.S. (1987). Evolution of a semi-Markov medium [in Russian]. *Cybernetics*, 5, 106–109.

Korolyuk, V.S. (1993). *Stochastic Models of Systems* [in Ukrainian]. Lybid, Kyiv.

Korolyuk, V.S. and Borovskikh, Y.V. (1981). *Analytical Problems for Asymptotics of Probability Distributions* [in Russian]. Naukova Dumka, Kyiv.

Korolyuk, V.S. and Korolyuk, V.V. (1999). *Stochastic Models of Systems*. Kluwer Academic, Dordrecht.

Korolyuk, V.S. and Limnios, N. (2004). Average and diffusion approximation for evolutionary systems in an asymptotic split phase space. *The Annals of Applied Probability*, 14(1), 489–516.

Korolyuk, V.S. and Limnios, N. (2005). *Stochastic Systems in Merging Phase Space*. World Scientific Publishing Company, New Jersey.

Korolyuk, V.S. and Limnios, N. (2009). Poisson approximation of processes with locally independent increments with Markov switching. *Theory of Stochastic Processes*, 15(1), 40–48.

Korolyuk, V.S. and Swishchuk, A.V. (1986). The central limit theorem for semi-Markov random evolutions. *Ukrainian Mathematical Journal*, 38, 286–289.

Korolyuk, V.S. and Swishchuk, A.V. (1995a). *Evolution of Systems in Random Media.* Chapman & Hall/CRC Press, London.

Korolyuk, V.S. and Swishchuk, A.V. (1995b). *Semi-Markov Random Evolution.* Kluwer Academic Publishers, Dordrecht.

Korolyuk, V.S. and Tadjiev, A. (1977). Asymptotic expansion for the distribution of absorption time of a semi-Markov process. *Reports of the USSR Academy of Sciences*, 12, 133–135.

Korolyuk, V.S. and Turbin, A.F. (1976). *Semi-Markov Processes and their Applications*. Naukova Dumka, Kyiv.

Korolyuk, V.S. and Turbin, A.F. (1982). *Markov Renewal Processes in Problems of Systems Reliability* [in Russian]. Naukova Dumka, Kyiv.

Korolyuk, V.S. and Turbin, A.F. (1993). *Mathematical Foundations of the State Lumping of Large Systems*. Kluwer Academic Publishers, Dordrecht.

Korolyuk, V.S., Penev, I.P., Turbin, A.F. (1973). Asymptotic expansions for the distribution of the absorption time of a Markov chain. *Cybernetics*, 4, 133–135.

Korolyuk, V.S., Turbin, A.F., Tomusyak, A.A. (1979a). The phase merging scheme of Markov chains based on phase lumping [in Russian]. In *Analytical Methods in the Theory of Probability*. Naukova Dumka, Kyiv.

Korolyuk, V.S., Turbin, A.F., Tomusyak, A.A. (1979b). Residence time of semi-Markov process in expanding set of states [in Russian]. In *Analytical Methods in the Theory of Probability*. Naukova Dumka, Kyiv.

Kovalenko, I.N. (1980). *Analysis of Rare Events in Evaluation of System Effectiveness and Reliability*. Sovetskoe Radio, Moscow.

Kovalenko, I.N., Yu., K.N., Shurenkov, V.M. (1983). *Random Processes: Manual.* Naukova Dumka, Kyiv.

Krein, S.G. (1971). Linear differential equations in banach space. *Proceedings of the American Mathematical Society, Translations of Mathematical Monographs*, 29, Rhode Island.

Kulkarni, V.G. (1997). Fluid models for single buffer systems. In *Frontiers in Queueing. Models and Applications in Science and Engineering*, Dshalalow, J.H. (ed.). CRC Press, Florida.

Lachal, A. (2006), Cyclic random motions in $\mathbb{R}^d$-space with n directions. *ESAIM: Probability and Statistics*, 10, 277–316.

Le, J.Y. and Raimond, O. (2004). Flows, coalescence and noise. *The Annals of Probability*, 32(2), 1247–1315.

Le Caër, G. (2010). A Pearson-Dirichlet random walk. *Journal of Statistical Physics*, 140, 728–751.

Le Caër, G. (2011). A new family of solvable Pearson-Dirichlet random walks. *Journal of Statistical Physics*, 144, 23–45.

Letac, G. and Piccioni, M. (2014). Dirichlet random walks. *Journal of Applied Probability*, 51, 1081–1099.

Lifshits, M. and Shi, Z. (2005). Aggregation rates in one-dimensional stochastic systems with adhesion and gravitation. *The Annals of Probability*, 33(1), 53–81.

Lindvall, T. (2002). *Lectures on the Coupling Method.* Dover Publications, New York.

López, O. and Ratanov, N. (2012). Option pricing under jump-telegraph model with random jumps. *Journal of Applied Probability*, 49(3), 838–849.

Maglaras, C. and Zeevi, A. (2004), Diffusion approximation for a multiclass Markovian service system with "guaranteed" and "best-effort" service level. *Mathematics of Operations Research*, 9(4), 786–813.

Malathornas, J.P., Perkins, J.D., Smith, R.L. (1983). The availability of a system of two unreliable machines connected by an intermediate storage tank. *AIIE Transactions*, 15(3), 195–201.

Masoliver, J., Porra, J.M., Weiss, G.H. (1993). Solution to the telegrapher's equation in the presence of reflecting and partly reflecting boundaries. *Physical Review*, 48(2), 939–944.

Mel'nichenko, I.P. and Plaksa, S.A. (2008). *Commutative Algebras and Spatial Potential Fields*. Institute of Mathematics NAS of Ukraine, Kyiv.

Meyn, S.P. and Tweedie, R.L. (1993). *Markov Chains and Stochastic Stability*. Springer-Verlag, New York.

Mitra, D. (1998). Stochastic theory of a fluid model of producers and consumers coupled by a buffer. *Advances in Applied Probability*, 20, 646–676.

Molchanov, S.A. and Yarovaya, E.B. (2013). Large deviations for a symmetric branching random walk on a multidimensional lattice. *Proceedings of the Steklov Institute of Mathematics*, 282, 186–201. Available at: https://doi.org/10.1134/S0081543813060163.

Nischenko, I.I. (2001). On asymptotic representation normalizing multiplier for random matrix-valued evolution. *Theory of Probability and Mathematical Statistics*, 64, 129–135.

Novak, S.Y. (2007). Measures of financial risks and market crashes. *Theory of Stochastic Processes*, 13(29), 182–193.

Novak, S.Y. (2011). *Extreme Value Methods with Applications to Finance*. Chapman & Hall/CRC Press, London.

Nummelin, E. (1984). *General Irreducible Markov Chain and the Non-negative Operators*. Cambridge University Press, New York.

Orsingher, E. (1985). Hyperbolic equations arising in random models. *Stochastic Processes and their Applications*, 21, 93–106.

Orsingher, E. and Beghin, L. (2009). Fractional diffusion equations and processes with randomly-varying time. *The Annals of Probability*, 37(1), 206–249.

Orsingher, E. and De Gregorio, A. (2007). Random flights in higher spaces. *Journal of Theoretical Probability*, 20, 769–806.

Orsingher, E. and Ratanov, N. (2002). Planar random motions with drift. *J. Appl. Math. Stoch.*, 15, 205–221.

Orsingher, E. and Somella, A.M. (2004). A cyclic random motion in $\mathbb{R}^3$ with four directions and finite velocity. *Stochastics and Stochastics Reports*, 76, 113–133.

Papanicolaou, G. (1971a). Asymptotic analysis of transport processes. *Bulletin of the American Mathematical Society*, 81, 330–391.

Papanicolaou, G. (1971b). Motion of a particle in a random field. *Journal of Mathematical Physics*, 12(24), 473–489.

Pinsky, M. (1991). *Lectures on Random Evolutions*. World Scientific, Singapore.

Pogorui, A.A. (1989). Distribution of reaching time of infinitely increasing level by family of semi-Markov processes with extending phase space. *Analytical Methods of Research of Evolution Stochastic Systems*. Institute of Mathematics of NAS of Ukraine, Kyiv.

Pogorui, A.A. (1990), Distribution of reaching time of infinitely increasing level by diffusion process on line. *Asymptotic and Applied Problems of Stochastic Evolution Theory*. Institute of Mathematics of NAS of Ukraine, Kyiv.

Pogorui, A.A. (1992). Asymptotic decomposition of distribution of reaching time of infinitely increasing level by semi-Markov process. *Stochastic Evolutions: Theoretical and Applied Problems*. Institute of Mathematics of NAS of Ukraine, Kyiv.

Pogorui, A.A. (1994). Asymptotic inequalities for the distribution of the time of stay of a semi-Markov process in an expanding set of states. *Ukrainian Mathematical Journal*, 46(11), 1757–1762.

Pogorui, A.A. (2003). Estimation of the stationary efficiency of a two-phase system with two storages. *Journal of Automation and Information Sciences*, 35, 16–23.

Pogorui, A.A. (2004a). Estimation of the efficiency of a data transmission line with feedback. *Journal of Automation and Information Sciences*, 36, 44–50.

Pogorui, A.A. (2004b). Reliability of two-phase transfer lines in parallel in Markov medium. *Int. Seminar Analisis: Norte-Sur, Mexico, Cinvestav, National Polytechnic Institute*, 96–97.

Pogorui, A.A. (2005). Stationary distribution of the random shift process with delaying in reflecting boundaries. *Scientific Journal of National Pedagogical Dragomanov University, Series 1*, 6, 168–172.

Pogorui, A.A. (2006). Stationary distribution of a process of random semi-Markov evolution with delaying screens in the case of balance. *Ukrainian Mathematical Journal*, 58(3), 430–437.

Pogorui, A.A. (2007). Hyperholomorphic functions in commutative algebras. *Complex Variables and Elliptic Equations*, 52(12), 1155–1159.

Pogorui, A.A. (2009a). Asymptotic expansion for distribution of Markovian random motion. *Random Operators and Stochastic Equations*, 17(2), 189–196.

Pogorui, A.A. (2009b). Stationary distributions of fading evolutions. *Ukrainian Mathematical Journal*, 61(3), 425–431.

Pogorui, A.A. (2010a). Asymptotic analysis for phase averaging of transport process. *Ukrainian Mathematical Journal*, 62(2), 190–198.

Pogorui, A.A. (2010b). Fading evolution in multidimensional spaces. *Ukrainian Mathematical Journal*, 62(11), 1577–1582.

Pogorui, A.A. (2011a). The distribution of random evolution in Erlang semi-Markov media. *Theory of Stochastic Processes*, 17(33), 90–99.

Pogorui, A.A. (2011b). Estimation of stationary productivity of one-phase system with a storage. *Markov Processes and Related Fields*, 17(2), 305–314.

Pogorui, A.A. (2012a). Evolution in multidimensional spaces. *Random Operators and Stochastic Equations*, 20(2), 135–141.

Pogorui, A.A. (2012b). System of interactive particles with Markovian switching. *Theory of Stochastic Processes*, 18(34–2), 83–95.

Pogorui, A.A. and Rodríguez-Dagnino, R.M. (2005a). One-dimensional semi-Markov evolution with general Erlang sojourn times. *Random Operators and Stochastic Equations*, 13(4), 399–405.

Pogorui, A.A. and Rodríguez-Dagnino, R.M. (2005b). Semi-Markov evolution with general Erlang sojourn times. *Int. Seminar: Analisis: Norte-Sur, Mexico, Cinvestav, National Polytechnic Institute*, 54–55.

Pogorui, A.A. and Rodríguez-Dagnino, R.M. (2006). Limiting distribution of random motion in an *n*-dimensional parallelepiped. *Random Operators and Stochastic Equations*, 14(4), 385–392.

Pogorui, A.A. and Rodríguez-Dagnino, R.M. (2008). Evolution process as an alternative to diffusion process and Black-Scholes formula. *5th Conference in Actuarial Science & Finance on Samos*, 4–7, Samos, Greece.

Pogorui, A.A. and Rodríguez-Dagnino, R.M. (2009a). Evolution process as an alternative to diffusion process and Black-Scholes formula. *Random Operators and Stochastic Equations*, 17(1), 61–68.

Pogorui, A.A. and Rodríguez-Dagnino, R.M. (2009b). Limiting distribution of fading evolution in some semi-Markov media. *Ukrainian Mathematical Journal*, 61(12), 1720–1724.

Pogorui, A.A. and Rodríguez-Dagnino, R.M. (2010a). Asymptotic expansion for transport processes in semi-Markov media. *Theory of Probability and Mathematical Statistics*, 83, 127–134.

Pogorui, A.A. and Rodríguez-Dagnino, R.M. (2010b). Stationary distribution of random motion with delay in reflecting boundaries. *Applied Mathematics*, 1(1), 24–28.

Pogorui, A.A. and Rodríguez-Dagnino, R.M. (2011a). Isotropic random motion at finite speed with k-Erlang distributed direction alternations. *Journal of Statistical Physics*, 145(1), 102–112.

Pogorui, A.A. and Rodríguez-Dagnino, R.M. (2011b). Multidimensional random motion with uniformly distributed changes of direction and Erlang steps. *Ukrainian Mathematical Journal*, 63(4), 572–577.

Pogorui, A.A. and Rodríguez-Dagnino, R.M. (2012). Random motion with uniformly distributed directions and random velocity. *Journal of Statistical Physics*, 147(6), 1216–1225.

Pogorui, A.A. and Rodríguez-Dagnino, R.M. (2013). Random motion with gamma steps in higher dimensions. *Statistics & Probability Letters*, 83, 1638–1643.

Pogorui, A.A. and Turbin, A.F. (2002). Estimation of stationary efficiency of a production line with two unreliable aggregates. *Cybernetics and Systems Analysis*, 38, 823–829.

Pogorui, A.A., Rodríguez-Said, R.D., Pogorui, A.A., Rodríguez-Dagnino, R.M. (2008). Stationary effectiveness of an information server with a single buffer and bursty demands of two different customers. *Stochastic Models*, 24, 246–269.

Pogorui, A.A., Rodríguez-Dagnino, R.M., Shapiro, M. (2014). Solutions for PDEs with constant coefficients and derivability of functions ranged in commutative algebras. *Mathematical Methods in the Applied Sciences*, 37(17), 2799–2810.

Qian, H., Raymond, G.M., Bassingthwaighte, J.B. (1998). On two-dimensional fractional Brownian motion and fractional Brownian random field. *Journal of Physics A: Mathematical and General*, 31, 1527–1535.

Ratanov, N. (2007). A jump telegraph model for option pricing. *Quantitative Finance*, 7(5), 575–583.

Ratanov, N. (2010). Option pricing model based on a Markov-modulated diffusion with jumps. *The Brazilian Journal of Probability and Statistics*, 24, 413–431.

Ratanov, N. and Melnikov, A. (2008). On financial markets based on telegraph processes. *Stochastics*, 80, 247–268.

Reed, M. and Simon, B. (1972). *Methods of Modern Mathematical Physics, Vol. 1.* Academic Press, New York.

Reibman, A., Smith, R., Trivedi, K.S. (1989). Markov and Markov reward model transient analysis: An overview of numerical approaches. *European Journal of Operational Research*, 40, 257–267.

Rodríguez-Said, R.D., Pogorui, A.A., Rodríguez-Dagnino, R.M. (2007). Stationary probability distribution of a system with n equal customers with bursty demands connected to a single buffer. *Random Operators and Stochastic Equations*, 15(2), 181–204.

Rodríguez-Said, R.D., Pogorui, A.A., Rodríguez-Dagnino, R.M. (2008). Stationary effectiveness of an information server with a single buffer and bursty demands of two different customers. *Stochastic Models*, 24, 246–269.

Samoilenko, I.V. (2001). Markovian random evolution in $\mathbb{R}^n$. *Random Operators and Stochastic Equations*, 9, 247–257.

Samoilenko, I.V. (2002). Fading Markov random evolution. *Ukrainian Mathematical Journal*, 54(3), 448–459.

Samoilenko, I.V. (2005). Asymptotic expansion for the functional of markovian evolution in $\mathbb{R}^d$ in the circuit of diffusion approximation. *Journal of Applied Mathematics and Stochastic Analysis*, 3, 247–257.

Sato, K. (1971). Potential operators for Markov processes. *Proceedings of the Sixth Berkeley Symposium on Math, Statistics and Probability*, 193–211, Berkeley, CA.

Semenov, A.T. (2008). Asymptotic expansions of the ruin probability for a generalized Poisson process. *Bulletin of Tomsk State University, Mathematics and Mechanics*, 3(4), 230–31.

Sericola, B. (1998). Transient analysis of stochastic fluid models. *Performance Evaluation*, 1(32), 245–263.

Sevastyanov, B.A. (1962). The effect of the bunker capacity on idle mean time of an automatic machine tool line problem [in Russian]. *Теория вероятностей и ее применения*, 7(4), 11–24.

Sevastyanov, B.A. (1971). *Branching Processes* [in Russian]. Nauka, Moscow.

Shurenkov, V.M. (1986). Markov intervention of chance and limit theorems. *Mathematics of the USSR-Sbornik*, 54, 1.

Shurenkov, V.M. (1989). *Ergodic Markov Processes* [in Russian]. Nauka, Moscow.

Silvestrov, D.S. (1980). *Semi-Markov Processes with a Discrete State Space. Library for an Engineer in Reliability*. Sovetskoe Radio, Moscow.

Silvestrov, D.S. (2004). *Limit Theorems for Randomly Stopped Stochastic Processes (Probability and Its Applications)*. Springer-Verlag, London.

Silvestrov, D.S. (2007a). Asymptotic expansions for distributions of the surplus prior and at the time of ruin. *Theory of Stochastic Processes*, 13(29), 183–88.

Silvestrov, D.S. (2007b). Asymptotic expansions for quasi-stationary distributions of nonlinearly perturbed semi-Markov processes. *Theory of Stochastic Processes*, 13(29), 267–271.

Sinai, Y. G. (1992). Distribution of some functionals of the integral of a random walk. *Theoretical and Mathematical Physics*, 90, 219–241.

Skorokhod, A.V. (1989). *Asymptotic Methods in the Theory of Stochastic Differential Equations*. American Mathematical Society, Rhode Island.

Sobolev, S.L. (1991). *Some Applications of Functional Analysis in Mathematical Physics*. American Mathematical Society, Rhode Island.

Soloviev, A.D. (1993). Asymptotic methods for highly reliable repairable systems. In *Handbook of Reliability Engineering*, Ushakov, I.A., Harrison, R. (eds). John Wiley & Sons, New York.

Stadje, W. (2007). The exact probability distribution of a two-dimensional random walk. *Journal of Statistical Physics*, 56, 207–216.

Stadje, W. and Zacks, S. (2004). Telegraph processes with random velocities. *Journal of Applied Probability*, 41, 665–678.

Stroock, D.W. and Varadhan, S.R.S. (1969). Diffusion processes with continuous coefficients. *Communications on Pure and Applied Mathematics*, 22(3), 345–400.

Stroock, D.W. and Varadhan, S.R.S. (1979). *Multidimensional Diffusion Processes*. Springer-Verlag, Berlin.

Sviridenko, M.N. (1989). Martingale approach to limit theorems for semi-Markov processes. *Theory of Probability and its Applications*, 34(3), 540–545.

Swishchuk, A.V. (1989). Weak convergence of semi-Markov random evolutions in an averaging scheme (martingale approach). *Ukrainian Mathematical Journal*, 41(3), 1450–1456.

Swishchuk, A.V. (1997). *Random Evolutions and their Applications*. Kluwer Academic Publishers, Dordrecht.

Swishchuk, A.V. (2000). *Random Evolutions and their Applications: New Trends*. Kluwer Academic Publishers, Dordrecht.

Swishchuk, A.V. (2004). Modeling of variance and volatility swaps for financial markets with stochastic volatilities. *Willmott Magazine*, 2, 64–72.

Swishchuk, A.V. (2020). *Inhomogeneous Random Evolutions and their Applications*. CRC Press, New York.

Swishchuk, A.V. and Burdeinyi, A.G. (1996). Stability of semi-Markov evolution systems and its application in financial mathematics. *Ukrainian Mathematical Journal*, 48(10), 1574–1591.

Swishchuk, A.V. and Islam, S. (2010). The geometric Markov renewal processes with applications to finance. *Stochastic Analysis and Applications*, 29(4), 684–705.

Turbin, A.F. (1972). The solution of some problems on perturbing the occurrence of matrices. *Reports of the National Academy of Sciences of Ukraine A*, 539–541.

Turbin, A.F. (1981). Limit theorems for perturbed semigroups and Markov processes in the scheme of asymptotic phase lumping [in Russian]. Preprint no. 80.18. *Institute of Mathematics, Ukrainian National Academy of Sciences*, Kyiv, 133–147.

Turbin, A.F. (1998). Mathematical models of one-dimensional Brownian motion as alternatives to the mathematical models of Einstein, Wiener and Levy [in Russian]. *Fractal Analysis and Related Fields*, 2, 47–60.

Vasileva, A.B. and Butuzov, V.F. (1973). *Asymptotic Expansion of Solutions of Singularly Perturbed Equations* [in Russian]. Nauka, Moscow.

Vasileva, A.B. and Butuzov, V.F. (1990). *Asymptotic Methods in the Theory of Singular Perturbation* [in Russian]. Vysshaya Shkola, Moscow.

Ventzel, A.D. and Freydlin, M.I. (1970). About small perturbations of dynamical systems. *Uspekhi Matematichaskikh Nauk.* [in Russian], 25(1), 3–55.

Vinogradov, O.P. (1994). On certain asymptotic properties of waiting time in a multiserver queueing system with identical times. *Theory of Probability and its Applications*, 39(4), 714–718.

Vishik, M. and Lusternik, L.A. (1960). The asymptotic behavior of solutions of linear differential equations with large or quickly changing coefficients and boundary conditions. *Russian Mathematical Surveys*, 15, 23–91.

Vysotsky, V.V. (2008). Clustering in a stochastic, model of one-dimensional gas. *The Annals of Applied Probability*, 18(3), 1026–1058.

Watkins, J. (1984). A central limit theorem in random evolution. *The Annals of Probability*, 12(2), 480–514.

Watkins, J. (1985). Limit theorems for stationary random evolutions. *Stochastic Processes and their Applications*, 19, 189–224.

Yeleyko, Y.I. and Zhernovyi, Y.V. (2002). Asymptotic properties of stochastic evolutions described by the solutions of the ordinary differential equations partially solved with respect to the higher derivatives. *Bulletin of the Lviv University, Series: Physics and Mathematics*, 60, 60–66.

Yu, P.A. (1999). The evolution of a system of particles and measure-valued processes. *Theory of Stochastic Processes*, 5(3–4), 188–197.

Yu, V.A. (2002). Coupling method for Markov chains under integral Doeblin type condition. *Theory of Stochastic Processes*, 8(24), 383–391.

Index

N, O

P, R

S

T, W

Summary of Volume 2

Chapter 5. Modeling and Pricing of Variance, Volatility, Covariance and Correlation Swaps for Financial Markets with Semi-Markov Volatilities

Other titles from

iSTE

in

Mathematics and Statistics

2020

BARBU Vlad Stefan, VERGNE Nicolas
Statistical Topics and Stochastic Models for Dependent Data with Applications

CHABANYUK Yaroslav, NIKITIN Anatolii, KHIMKA Uliana
Asymptotic Analyses for Complex Evolutionary Systems with Markov and Semi-Markov Switching Using Approximation Schemes

KOROLIOUK Dmitri
Dynamics of Statistical Experiments

MANOU-ABI Solym Mawaki, DABO-NIANG Sophie, SALONE Jean-Jacques
Mathematical Modeling of Random and Deterministic Phenomena

2019

BANNA Oksana, MISHURA Yuliya, RALCHENKO Kostiantyn, SHKLYAR Sergiy
Fractional Brownian Motion: Approximations and Projections

GANA Kamel, BROC Guillaume
Structural Equation Modeling with lavaan

KUKUSH Alexander
Gaussian Measures in Hilbert Space: Construction and Properties

LUZ Maksym, MOKLYACHUK Mikhail
Estimation of Stochastic Processes with Stationary Increments and Cointegrated Sequences

MICHELITSCH Thomas, PÉREZ RIASCOS Alejandro, COLLET Bernard, NOWAKOWSKI Andrzej, NICOLLEAU Franck
Fractional Dynamics on Networks and Lattices

VOTSI Irene, LIMNIOS Nikolaos, PAPADIMITRIOU Eleftheria, TSAKLIDIS George
Earthquake Statistical Analysis through Multi-state Modeling
(Statistical Methods for Earthquakes Set – Volume 2)

2018

AZAÏS Romain, BOUGUET Florian
Statistical Inference for Piecewise-deterministic Markov Processes

IBRAHIMI Mohammed
Mergers & Acquisitions: Theory, Strategy, Finance

PARROCHIA Daniel
Mathematics and Philosophy

2017

CARONI Chysseis
First Hitting Time Regression Models: Lifetime Data Analysis Based on Underlying Stochastic Processes
(Mathematical Models and Methods in Reliability Set – Volume 4)

CELANT Giorgio, BRONIATOWSKI Michel
Interpolation and Extrapolation Optimal Designs 2: Finite Dimensional General Models

CONSOLE Rodolfo, MURRU Maura, FALCONE Giuseppe
Earthquake Occurrence: Short- and Long-term Models and their Validation
(Statistical Methods for Earthquakes Set – Volume 1)

D'AMICO Guglielmo, DI BIASE Giuseppe, JANSSEN Jacques, MANCA Raimondo
Semi-Markov Migration Models for Credit Risk
(Stochastic Models for Insurance Set – Volume 1)

GONZÁLEZ VELASCO Miguel, del PUERTO GARCÍA Inés, YANEV George P.
Controlled Branching Processes
(Branching Processes, Branching Random Walks and Branching Particle Fields Set – Volume 2)

HARLAMOV Boris
Stochastic Analysis of Risk and Management
(Stochastic Models in Survival Analysis and Reliability Set – Volume 2)

KERSTING Götz, VATUTIN Vladimir
Discrete Time Branching Processes in Random Environment
(Branching Processes, Branching Random Walks and Branching Particle Fields Set – Volume 1)

MISHURA YULIYA, SHEVCHENKO Georgiy
Theory and Statistical Applications of Stochastic Processes

NIKULIN Mikhail, CHIMITOVA Ekaterina
Chi-squared Goodness-of-fit Tests for Censored Data
(Stochastic Models in Survival Analysis and Reliability Set – Volume 3)

SIMON Jacques
Banach, Fréchet, Hilbert and Neumann Spaces
(Analysis for PDEs Set – Volume 1)

2016

CELANT Giorgio, BRONIATOWSKI Michel
Interpolation and Extrapolation Optimal Designs 1: Polynomial Regression and Approximation Theory

CHIASSERINI Carla Fabiana, GRIBAUDO Marco, MANINI Daniele
Analytical Modeling of Wireless Communication Systems
(Stochastic Models in Computer Science and Telecommunication Networks Set – Volume 1)

GOUDON Thierry
Mathematics for Modeling and Scientific Computing

KAHLE Waltraud, MERCIER Sophie, PAROISSIN Christian
Degradation Processes in Reliability
(Mathematial Models and Methods in Reliability Set – Volume 3)

KERN Michel
Numerical Methods for Inverse Problems

RYKOV Vladimir
Reliability of Engineering Systems and Technological Risks
(Stochastic Models in Survival Analysis and Reliability Set – Volume 1)

2015

DE SAPORTA Benoîte, DUFOUR François, ZHANG Huilong
Numerical Methods for Simulation and Optimization of Piecewise Deterministic Markov Processes

DEVOLDER Pierre, JANSSEN Jacques, MANCA Raimondo
Basic Stochastic Processes

LE GAT Yves
Recurrent Event Modeling Based on the Yule Process
(Mathematical Models and Methods in Reliability Set – Volume 2)

2014

COOKE Roger M., NIEBOER Daan, MISIEWICZ Jolanta
Fat-tailed Distributions: Data, Diagnostics and Dependence
(Mathematical Models and Methods in Reliability Set – Volume 1)

MACKEVIČIUS Vigirdas
Integral and Measure: From Rather Simple to Rather Complex

PASCHOS Vangelis Th
Combinatorial Optimization – 3-volume series – 2nd edition
Concepts of Combinatorial Optimization / Concepts and Fundamentals – volume 1
Paradigms of Combinatorial Optimization – volume 2
Applications of Combinatorial Optimization – volume 3

2013

COUALLIER Vincent, GERVILLE-RÉACHE Léo, HUBER Catherine, LIMNIOS Nikolaos, MESBAH Mounir
Statistical Models and Methods for Reliability and Survival Analysis

JANSSEN Jacques, MANCA Oronzio, MANCA Raimondo
Applied Diffusion Processes from Engineering to Finance

SERICOLA Bruno
Markov Chains: Theory, Algorithms and Applications

2012

BOSQ Denis
Mathematical Statistics and Stochastic Processes

CHRISTENSEN Karl Bang, KREINER Svend, MESBAH Mounir
Rasch Models in Health

DEVOLDER Pierre, JANSSEN Jacques, MANCA Raimondo
Stochastic Methods for Pension Funds

2011

MACKEVIČIUS Vigirdas
Introduction to Stochastic Analysis: Integrals and Differential Equations

MAHJOUB Ridha

Recent Progress in Combinatorial Optimization – ISCO2010

RAYNAUD Hervé, ARROW Kenneth
Managerial Logic

2010

BAGDONAVIČIUS Vilijandas, KRUOPIS Julius, NIKULIN Mikhail
Nonparametric Tests for Censored Data

BAGDONAVIČIUS Vilijandas, KRUOPIS Julius, NIKULIN Mikhail
Nonparametric Tests for Complete Data

IOSIFESCU Marius *et al.*
Introduction to Stochastic Models

VASSILIOU PCG
Discrete-time Asset Pricing Models in Applied Stochastic Finance

2008

ANISIMOV Vladimir
Switching Processes in Queuing Models

FICHE Georges, HÉBUTERNE Gérard
Mathematics for Engineers

HUBER Catherine, LIMNIOS Nikolaos *et al.*
Mathematical Methods in Survival Analysis, Reliability and Quality of Life

JANSSEN Jacques, MANCA Raimondo, VOLPE Ernesto
Mathematical Finance

2007

HARLAMOV Boris
Continuous Semi-Markov Processes

2006

CLERC Maurice
Particle Swarm Optimization